Small-Bore Heating and Hot-Water Supply for Small Dwellings

Small-Bore Heating and Hot-Water Supply for Small Dwellings

J. J. BARTON
F.I.H.V.E., M.INST.F., HON.I.D.H.E.
(Heating and Ventilating Consulting Engineer)

LONDON
NEWNES - BUTTERWORTHS

ENGLAND:	Butterworth & Co. (Publishers) Ltd. London: 88 Kingsway, WC2B 6AB
AUSTRALIA:	Butterworth & Co. (Australia) Ltd. Sydney: 20 Loftus Street Melbourne: 343 Little Collins Street Brisbane: 240 Queen Street
CANADA:	Butterworth & Co. (Canada) Ltd. Toronto: 14 Curity Avenue, 374
NEW ZEALAND:	Butterworth & Co. (New Zealand) Ltd. Wellington: 49/51 Ballance Street Auckland: 35 High Street
SOUTH AFRICA:	Butterworth & Co. (South Africa) (Pty.) Ltd. Durban: 33/35 Beach Grove

Published in 1964 by Newnes-Butterworths,
an imprint of the Butterworth Group

Reprinted 1969
Second edition 1970

ISBN 0 408 00017 1

Filmset by Filmtype Services Ltd, Scarborough, Yorkshire
Printed by photo-lithography at the Pitman Press, Bath

Preface

The preparation of this second edition has involved extensive revision of the original work. The scope of the book has been extended to include two-pipe systems and the use of 10 mm ($\frac{3}{8}$ in) and 6 mm ($\frac{1}{4}$ in) bore sub-circuits.

The lack of information on radiator connection sizing in the first edition has now been remedied by the inclusion of a new chapter dealing fully with this important subject.

During the period 1971-72, the change to the production of drawings and documents in metric units for all new contracts in the construction industry will take place. Accordingly, in this edition, the SI system of measurement and decimal currency have superseded their Imperial counterparts.

To ensure the usefulness of the book as a reference work, especially during the changeover to the metric system, the appendices have been extended to include, where necessary, essential tabular design information in both SI and Imperial units, and also notes on decimal currency.

In revising the book, I have tried to retain the practical nature of the original work, which I hope will not only meet the needs of all those designers and contractors directly concerned with domestic central heating, but will also help young students and apprentices in the plumbing, heating and building industries, and also students of domestic architecture.

The work conforms with the requirements of British Standard Code of Practice, CP 3006: Part 1: 1969 (copies of which are available from the BSI Sales Office, 101 Pentonville Road, London N1), and 'The Small Bore Domestic Central Heating Guide to Good Practice', prepared jointly by the Institution of Heating and Ventilating Engineers, 49 Cadogan Square, London SW1, and the Heating and Ventilating Contractors Association, Coastal Chambers, 172 Buckingham Palace Road, London SW1, from whom copies are obtainable.

Southport J. J. Barton

Contents

Appendices

1

Estimating the Heat Loss

The first step in the design of a space-heating system is to estimate the heat requirements of the building. When the building is maintained at a temperature above that of the outside air, heat will flow from the rooms through the structure to the outside atmosphere. The heating system must replace this heat loss and also provide heat to make good that lost by air change. Normal air change may be due to infiltration which occurs in every building and partly to planned ventilation where this is installed.

Design Temperatures

Space heating in this country is designed on a basic outside temperature of −1°C. Recommended room temperatures and air-change rates for residences are given in Table 1.

Where background heating only is required, the design temperatures of living rooms may be reduced provided some form of topping-up heat is available for use during the colder periods of the heating season. The central heating in the lounge for example may be designed to keep the temperature at 16°C with −1°C outside, the topping-up heat to bring the room to 21°C being furnished by an open fire, a gas fire, or one of the many forms of electric heaters. This also applies to the dining room.

Entrance halls, corridors and landings should always be maintained at the temperatures recommended in Table 1. A warm entrance hall, corridor and landing have a profoundly good influence upon the comfort of living rooms and bedrooms, and background heating is not advised in these parts of the house. Those bathrooms in which the hot water cylinder is sited in an airing cupboard within the room, may with safety be background heated up to 13°C as the heat from the airing cupboard will top up to 16°C–18°C. The great majority of people, however, like their bathroom really warm, and the stray heat provided by a hot water cylinder is generally welcome even when the bathroom heating has been designed to give full warmth.

Table 1
Recommended room air temperatures and air-change rates for dwellings *

Name of room	*Air temperature*		*Air changes per hour (natural ventilation)*
	°C	*°F*	
Living room, dining room, bed-sitting room	21	70	2
Bathroom	18	65	2
Hall, staircase, landing, corridor, kitchen	16	60	2
Bedroom	13	55	2
Toilet	16	60	1

*Mainly as recommended in British Standard Code of Practice CP 3006 : Part 1 : 1969
Column 2 gives converted values of Fahrenheit temperatures (col. 3) from the equation: °C = (°F − 32)/1·8, and rounded off to the nearest whole number.

In those cases where the hot-water cylinder is not sited in the bathroom, the design temperature should be raised to 18°C.

The basic metric unit of energy, or 'quantity of heat', is the kilocalorie (kcal), and the SI unit of heat quantity is the joule (J).

1 kcal = 4187 J and
1 Btu = 1055 J

The SI equivalent of the Imperial unit of heat flow (Btu/h) is the joule per second (J/s), and 1 J/s = 1 watt, which is the term used to express SI heat flow.

Btu/h × 0·2931 = watts, which is abbreviated W, except in calculations involving the mass flow rate *W*, when to avoid confusion watts is written in full.

Estimating the Heat Loss

The actual amount of heat lost through walls, floors, roof and windows, together with that lost by air change, must be calculated, and to do this a knowledge of the thermal transmittance of the various parts of the structure is necessary.

Thermal transmittance can be defined as 'the amount of heat in watts (W) which will flow through $1 m^2$ of a particular type and thickness of construction when the temperatures of the air adjoining the inner and outer face differ by 1°C.' This is called the '*U* value,' and for practical purposes *U* values have been determined experimentally by research bodies and engineers. Complete tables have been published under the auspices of the British Standards Institution, and are accepted by the Institution of Heating and Ventilating Engineers as the standard values (IHVE Guide 1965). The *U* values in Table 2 cover the structural elements from which traditional dwellings in this country are built. The values have been converted from standard Imperial units, using the conversion factor

$$U = \text{Btu/h ft}^2\,{}^\circ\text{F} \times 5{\cdot}678 = \text{W/m}^2\,{}^\circ\text{C}$$

which is the heat transmitted in watts (joules per second) through an area of one square metre ($1 m^2$) of a building element (*viz:* wall, floor, roof etc.). The heat transmission *H* through a structure is given by

$$H = U\,(t_i - t_o) \times A$$

Where *U* is the transmission coefficient of the structure, W/m^2 °C (Btu/h ft^2 °F)

t_i is the inside air temperature
t_o is the outside air temperature
A is the area of the structure

(For Imperial *U* coefficients see Table 1 Appendix A).

Houses having a sheltered exposure to the wind are those in the interior of towns. Normal exposure applies to suburban or country dwellings. A house built upon an exposed hill site, or by the coast or riverside will have a severe exposure.

The examples which follow serve to show the use of the tables.

Example 1

A 279 mm unventilated cavity south wall measures 5·18 m long × 2·44 m high and has one window which is 2·44 × 1·37 m.

Calculate the heat loss from the actual wall surface.

Room temperature 16°C, outside temperature −1°C. Exposure to wind: normal.

Solution

From Table 2 the *U* value is 1·59 W/m²°C.

Area of wall	5·18 × 2·44 m	= 12·63
Area of window	2·44 × 1·37 m	= 3·35
Net area of wall		= 9·28 m²

$$\text{Heat loss} = 1{\cdot}59 \times 9{\cdot}28 \times 16-(-1)$$
$$= 1{\cdot}59 \times 9{\cdot}28 \times 17$$
$$= 250{\cdot}7 \simeq 251 \text{ W}$$

Example 2

Calculate the heat loss through a living-room ceiling to the bedroom above. Floor construction is wood on joists with plaster ceiling.

Living-room temperature 21°C, bedroom 13°C.

Solution

From Table 2, *U* = 1·65 W/m²°C.

Ceiling dimensions: 4·57 m × 3·01 m = 13·76 m²

Heat flow upwards to bedroom = 1·65 × 13·76 × (21–13)

= 1·65 × 13·76 × 8

= 182 W

Heat Loss Due to Air Change

In SI terms the specific heat capacity of air (volume basis) is 1·34 kilojoules per cubic metre per degree C, or 1·34kJ/m³°C, which means that each cubic metre of air admitted to a room requires 1·34 kJ to raise its temperature by 1°C

An example of the calculation to determine heat loss due to air change follows.

Example 3

Calculate the heat loss due to air change for a living room 4·57 m long × 3·01 m wide × 2·44 m high, having two external walls with windows.

Room temperature 21°C. Outside temperature − 1°C.

Solution

From Table 1, air change = 2/h.

Volume of air to be heated will be

$$V = (4{\cdot}57\text{ m} \times 3{\cdot}01\text{ m} \times 2{\cdot}44\text{ m}) \times 2$$
$$= 33{\cdot}50 \times 2$$
$$= 67\text{ m}^3$$

The temperature of the air entering the room is -1°C, and as the room temperature is 21°C, we must determine the quantity of heat required to raise the temperature of 67 m^3 of air in one hour.

As stated earlier the specific heat capacity of air (volume basis) is 1·34 kJ/m^3°C, and the hourly air heat loss in this case will be

$$\text{Air heat loss} = 67\text{ m}^3 \times 1{\cdot}34 \times 21 - (-1)\text{ kJ/h}$$
$$= 67 \times 1{\cdot}34 \times 22$$
$$= 1975\text{ kJ/h.}$$

To give a result in SI units (kJ/s or kW), it is necessary to divide by 3600

$$\text{Air heat loss} = \frac{V \times AC \times 1{\cdot}34 \times \Delta t}{3600}\text{ kW}$$

Where V = the volume of room in cubic metres
AC = the number of air changes per hour
Δt = the temperature difference in degrees C.

Substituting in the present example, we have

$$\text{Air heat loss} = \frac{V \times AC \times 1{\cdot}34 \times \Delta t}{3600}$$
$$= \frac{33{\cdot}5 \times 2 \times 1{\cdot}34 \times 22}{3600}$$
$$= \frac{1975}{3600^*}$$
$$= 0{\cdot}549\text{ kW or 549 W}$$

*When the answer is required directly in watts, divide by 3·600.

Table 2
Heat transmittance coefficients for the common structural components of domestic dwelling houses ($U = W/m^{2}{}^{\circ}C$)

Orientation	*Exposure to wind*					
S	*Sheltered*	*Normal*	*Severe*			
W, SW, SE		*Sheltered*	*Normal*	*Severe*		
NW			*Sheltered*	*Normal*	*Severe*	
N, NE, E			*Sheltered*	*Normal*		*Severe*
Brick walls						
114 mm brick, plastered one side	2·61	2·78	3·01	3·24	3·46	3·69
229 mm brick, plastered one side	2·04	2·16	2·33	2·44	2·56	2·73
279 mm cavity brick, plastered one side (unventilated)	1·53	1·59	1·65	1·70	1·76	1·82
279 mm cavity brick, plastered one side (ventilated)	1·70	1·76	1·87	1·93	2·04	2·10
Windows						
Single glazing	3·98	4·49	5·00	5·68	6·47	7·38
Double glazing	2·33	2·50	2·67	2·84	3·01	3·18

Ground floors	*U for inside-outside temperature difference*
(1) Solid concrete in contact with earth	1·14
(2) As (1) with fitted carpet and underlay	0·91
(3) Ventilated wood floor on joists, air bricks on one side, bare boards	1·70
(4) As (3) with fitted carpet and underlay	1·25
(5) As (3) with air bricks on more than one side	2·27
(6) As (5) with fitted carpet and underlay	1·53

Intermediate floors	*Heat flow downwards*	*Heat flow upwards*
(1) Wood on joists, plaster ceiling	1·25	1·65
(2) As (1) with fitted carpet and underlay	0·97	1·19

	Sheltered	Normal	Severe
Roofs (flat)			
(1) 102 mm concrete, asphalted, with plaster ceiling	3·47	3·80	4·15
(2) As (1) with 25 mm glass fibre, or mineral wool board under asphalt	1·02	1·14	1·25
Roofs (pitched)			
(1) Tiles on battens, plaster ceiling; roof space	2·84	3·18	3·63
(2) As (1) with 25 mm glass fibre or mineral wool insulation between or over ceiling joists	0·97	1·02	1·14

The Effect of Thermally Insulating the Building

Whether the medium chosen to heat a house is oil, solid fuel, electricity or gas, good structural insulation will reduce both capital and running costs considerably.

Insulating the roof with 51 mm thick mineral wool, or glass-fibre quilts will in most cases give a U value of 0·85 W/m^2°C and result in an annual saving in fuel of the order of 12 to 15 per cent. It can be shown that where the roof of a house is insulated to a U value of 0·85, and the outside walls to a U value of 1·14, central heating in bedrooms can be entirely dispensed with, provided that the ground-floor rooms are centrally heated to 21°C.

Insulating slabs having a thermal conductivity of 0·036 W/m °C at normal building temperatures applied between the inner and outer layers of cavity brick walls give the following results:

13 mm insulating slabs in 279 mm walls: $U = 1{\cdot}08$
25 mm insulating slabs in 279 mm walls: $U = 0{\cdot}79$

Figs. 1A and 1B represent the ground- and first-floor plans of a standard type of three-bedroom semi-detached house.

Using the tables and information already given, the heat losses from this building can be calculated. For calculation purposes it will be assumed that the adjacent house is not centrally heated and that consequently there will be some heat loss through the north wall. In cases of this kind a temperature difference of 6°C is allowed when calculating heat loss through party walls.

Air-change rates, U values and temperatures given in Table 3 have been abstracted from Tables 1 and 2, and will be used in the calculations which follow. The basic details upon which the heat losses are to be calculated are as follows:

Outside temperature = -1°C.
Room temperatures, see Table 3.
U values, see Table 3.
Air-change rates, see Table 3.
Outside walls are 279 mm cavity unventilated, and plastered.
Inside walls 114 mm plastered.
Party wall 229 mm plastered.
First floor: wood on joists, plaster ceiling.
Ground floor: solid concrete in contact with earth.
Roof: pitched, tiles on battens, plaster ceiling with roof space.

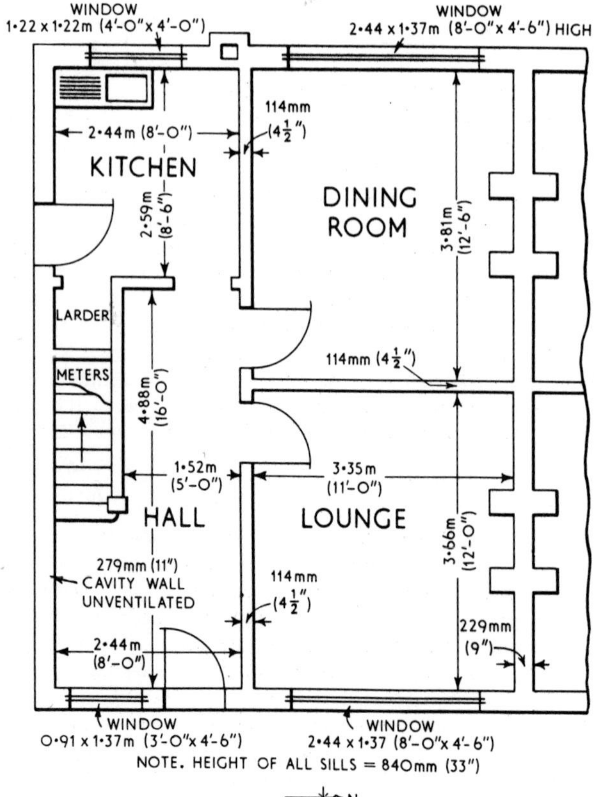

Fig. 1A *Ground-floor plan*

Height of rooms: 2·44 m from floor to ceiling.

Reference to the plans Figs. 1A and 1B and the room temperatures, Table 3, indicates that owing to the temperature difference between certain rooms heat will flow from the higher to the lower heated spaces. For example, upward flow will take place from the ground floor rooms to the bedrooms, and through internal walls on both floors. This heat gain although small in quantity is advantageous to the general comfort condition of a centrally

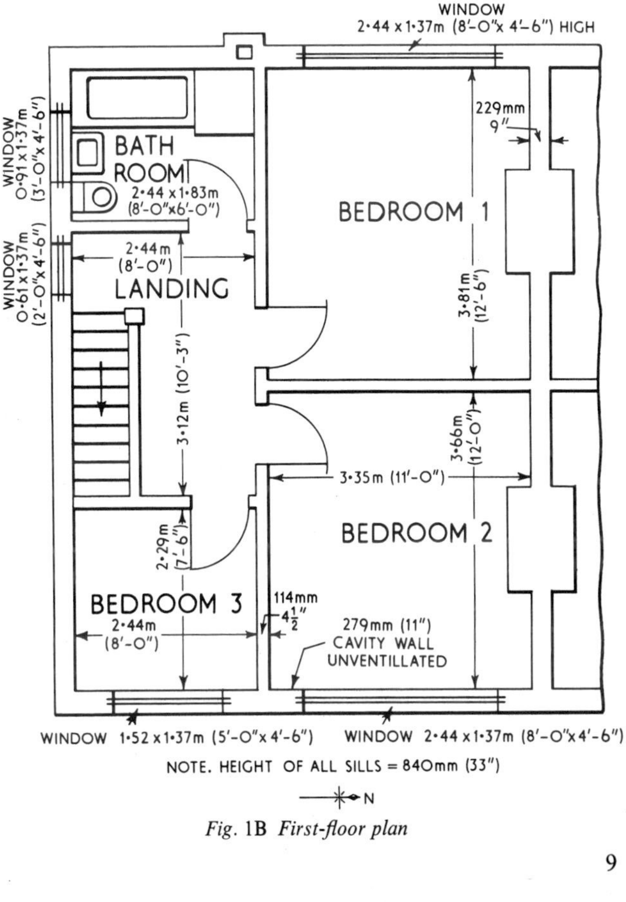

Fig. 1B *First-floor plan*

Table 3

Air-change rates, *U* factors ($W/m^{2}{}^{\circ}C$), and room temperatures for typical three-bedroom semi-detached house in Figs. 1A and 1B

Name of room	*Temperature °C*	*Air change per hour*	*U factors and orientation of surfaces*				
			Glass	*Wall*	*Roof*	*Ground floor*	*First floor*
Dining room	21	2	W 5·00	W 1·65 S 2·78 N 2·44		1·14	1·65
Lounge	21	2	E 5·68	E 1·70 S 2·78 N 2·44		1·14	1·65
Hall	16	2	E 5·68	E 1·70 S 1·59		1·14	1·65
Bathroom	16	2	S 4·49	W 1·65 S 1·59 N 3·24	3·17		
Bedroom 1	13	2	W 5·00	W 1·65 N 2·44	3·17		
Bedroom 2	13	2	E 5·68	E 1·70 N 2·44	3·17		
Bedroom 3	13	2	E 5·68	E 1·70 S 1·59	3·17		

heated dwelling, especially for bedrooms which are usually kept slightly above their relatively low design temperature. In fact during the greater part of the heating season the bedroom radiators may be shut off, and the design temperature will still be maintained.

Heat-Loss Calculations

Dining Room

Length 3·81 m, width 3·35 m, height 2·44 m

Room cube = 31·14 m^3

		watts
Air heat loss	$= \dfrac{31{\cdot}14 \times 1{\cdot}34 \times 2AC \times 22}{3{\cdot}600}$	= 508
Ground-floor heat loss	= 12·73 m^2 × 1·14 × 22	= 319
First-floor (upwards) heat loss	= 12·73 m^2 × 1·65 × (21 − 13)	= 168
Window heat loss	= 3·35 m^2 × 5·00 × 22	= 369
West-wall heat loss	= 4·83 m^2 × 1·65 × 22	= 175
South-wall heat loss	= 9·29 m^2 × 2·78 × (21 − 16)	= 129
North-wall heat loss*	= 9·29 m^2 × 2·44 × (21 − 16)	= 113
Heat loss from dining room		= 1781 W

*Where it is known that the adjoining house is not centrally heated the loss through this wall must be included.

Lounge

Length 3·66 m, width 3·35 m, height 2·44 m

Room cube = 29·9 m^3

		watts
Air heat loss	$= \dfrac{29{\cdot}9 \times 1{\cdot}34 \times 2AC \times 22}{3{\cdot}600}$	= 490
Ground-floor heat loss	= 12·26 m^2 × 1·14 × 22	= 308
First-floor (upwards) heat loss	= 12·26 m^2 × 1·65 × (21 − 13)	= 162
Window heat loss	= 3·35 m^2 × 5·68 × 22	= 418

East-wall heat loss	$= 4{\cdot}83\ m^2 \times 1{\cdot}70 \times 22$	=	181
South-wall heat loss	$= 8{\cdot}92\ m^2 \times 2{\cdot}78 \times (21 - 16)$	=	124
North-wall heat loss*	$= 8{\cdot}92\ m^2 \times 2{\cdot}44 \times (21 - 16)$	=	109
Heat loss from lounge		=	1792 W

*Heat loss to adjoining house

Hall

Length 2·44 m, width 2·13 m, height 2·44 m
Length 2·74 m, width 1·52 m, height 2·44 m
Room cube = 22·88 m^3

			watts
Air heat loss	$= \dfrac{22{\cdot}88 \times 1{\cdot}34 \times 2AC \times 17}{3{\cdot}600}$	=	290
Ground-floor heat loss	$= 9{\cdot}38\ m^2 \times 1{\cdot}14 \times 17$	=	182
First-floor (upwards) heat loss	$= 5{\cdot}57\ m^2 \times 1{\cdot}65 \times (16 - 13)$	=	28
Window and glazed door heat loss	$= 1{\cdot}53\ m^2 \times 5{\cdot}68 \times 17$	=	148
East-wall heat loss	$= 4{\cdot}41\ m^2 \times 1{\cdot}70 \times 17$	=	128
South-wall heat loss	$= 11{\cdot}89\ m^2 \times 1{\cdot}59 \times 17$	=	321
Heat loss from hall		=	1097W

Note: owing to the irregular shape of the hall, two sets of measurements are taken to establish the room cube. This also applies to the floor area.

Bathroom

Length 2·44 m, width 1·83 m, height 2·44 m
Room cube, less cylinder cupboard = 9·06 m^3

			watts
Air heat loss	$= \dfrac{9{\cdot}06 \times 1{\cdot}34 \times 2AC}{3{\cdot}600}$	=	115
Floor–No heat loss			
Roof heat loss	$= 3{\cdot}72\ m^2 \times 3{\cdot}17 \times 17$	=	200

Window heat loss	$= 1{\cdot}26\ m^2 \times 4{\cdot}49 \times 17$	=	96
South-wall heat loss	$= 3{\cdot}21\ m^2 \times 1{\cdot}59 \times 17$	=	87
West-wall heat loss	$= 3{\cdot}90\ m^2 \times 1{\cdot}65 \times 17$	=	108
North-wall heat loss	$= 2{\cdot}42\ m^2 \times 3{\cdot}24 \times (16 - 13)$	=	24
Bathroom heat loss		=	630W

First-Floor Landing

Length 3·12 m, width 2·44 m, height 2·44 m

Room cube = 18·58 m³ *watts*

$$\text{Air heat loss} = \frac{18{\cdot}58 \times 1{\cdot}34 \times 2AC \times 17}{3{\cdot}600} = 235$$

Floor–No heat loss			
Roof heat loss	$= 7{\cdot}62\ m^2 \times 3{\cdot}17 \times 17$	=	411
Window heat loss	$= 0{\cdot}84\ m^2 \times 4{\cdot}49 \times 17$	=	64
South-wall heat loss	$= 6{\cdot}78\ m^2 \times 1{\cdot}59 \times 17$	=	183
North-wall heat loss	$= 7{\cdot}62\ m^2 \times 3{\cdot}24 \times 17$	=	74
East-wall heat loss	$= 5{\cdot}95\ m^2 \times 3{\cdot}24 \times (16 - 13)$	=	58
Heat loss from first-floor landing		=	1025W

Bedroom 1

Length 3·81 m, width 3·35 m, height 2·44 m

Room cube = 31·15 m³ *watts*

$$\text{Air heat loss} = \frac{31{\cdot}15 \times 1{\cdot}34 \times 2AC \times 14}{3{\cdot}600} = 324$$

Floor–No heat loss			
Roof heat loss	$= 12{\cdot}73\ m^2 \times 3{\cdot}17 \times 14$	=	565
Window heat loss	$= 3{\cdot}35\ m^2 \times 5{\cdot}00 \times 14$	=	234
West-hall heat loss	$= 4{\cdot}83\ m^2 \times 1{\cdot}65 \times 14$	=	112
North-wall heat loss	$= 9{\cdot}29\ m^2 \times 2{\cdot}44 \times 6$	=	136
Heat loss from bedroom 1		=	1371W

Bedroom 2

Length 3·66 m, width 3·35 m, height 2·44 m
Room cube = 29·9 m³ *watts*

$$\text{Air heat loss} = \frac{29{\cdot}9 \times 1{\cdot}34 \times 2AC \times 14}{3{\cdot}600} = 312$$

		watts
Floor–No heat loss		
Roof heat loss	$= 12{\cdot}26\ m^2 \times 3{\cdot}17 \times 14$	= 544
Window heat loss	$= 3{\cdot}35\ m^2 \times 5{\cdot}68 \times 14$	= 266
East-wall heat loss	$= 4{\cdot}83\ m^2 \times 1{\cdot}70 \times 14$	= 115
North-wall heat loss	$= 8{\cdot}92\ m^2 \times 2{\cdot}44 \times (13 - 7)$	= 130
Heat loss from bedroom 2		= 1367W

Bedroom 3

Length 2·44 m, width 2·29 m, height 2·44 m
Room cube = 13·59 m³ *watts*

$$\text{Air heat loss} = \frac{13{\cdot}59 \times 1{\cdot}34 \times 2\,AC \times 14}{3{\cdot}600} = 142$$

		watts
Floor–No heat loss		
Roof heat loss	$= 5{\cdot}57\ m^2 \times 3{\cdot}17 \times 14$	= 247
Window heat loss	$= 2{\cdot}09\ m^2 \times 5{\cdot}68 \times 14$	= 166
East-wall heat loss	$= 3{\cdot}86\ m^2 \times 1{\cdot}70 \times 14$	= 92
South-wall heat loss	$= 5{\cdot}57\ m^2 \times 1{\cdot}59 \times 14$	= 124
Heat loss from bedroom 3		= 771W

Note on Air Change in Living Rooms

A very important aspect of air change in a room heated by a combination of central heating and an open fireplace should be considered at this stage.

The above figures show that the heat loss due to just one air change represents 15·5 per cent of the total estimated room heat loss. If the central heating is to succeed, steps must be taken to ensure that the air-change rate is not exceeded. With the older type of fireplace a considerable amount of air in excess of that required for combustion is drawn up the chimney.

Where central heating is installed in such rooms the chimney pull should be restricted to near the value necessary for good combustion only. Modern fireplaces are now available with adjustable restrictive throats which enable the flow of air to the chimney to be kept within the required limits.

In the present example the boiler will be sited in the kitchen and will (as shown in Chapter 6) maintain the room at a satisfactory temperature. In cases where the boiler is not placed in the kitchen, heat losses should be calculated for this space on the basis of 16°C room temperature with two air changes per hour.

Summary of Heat Losses	*watts*
Dining room	1781
Lounge	1792
Hall	1097
Bathroom	630
Landing	1025
Bedroom 1	1371
Bedroom 2	1367
Bedroom 3	771
Total building heat loss	= 9834W

Rule-of-Thumb Methods of Design

House heating installations designed by rule-of-thumb methods are usually unsatisfactory in that they under-heat or over-heat the building. Careful measurement of the building and close estimation of the basic heat requirements are the first essentials. No matter how good the pipe layout and sizing, or how well the pump and boiler have been sized, if the heat losses are not estimated properly the heating installation will rarely give the efficient performance which the customer is entitled to expect.

When one considers the many variable factors involved such as the thickness and nature of the structural elements, window surfaces in relation to wall areas, and cube of room; orientation, exposure to wind, and temperature differences between rooms, it becomes unwise to base heat loss upon some hit-or-miss value of so many watts per cubic metre of space, or per square metre of floor area.

Packaged House-Heating Systems

This modern development has obvious advantages but if used indiscriminately can only bring disappointment for the customer and discredit upon the installer.

The principle of the packaged unit for house heating can be successfully applied only where a number of houses of exactly similar dimensions, design, orientation and exposure to wind, are to be heated. A builder developing a normally exposed site upon which he plans to build say fifty pairs of houses as Figs. 1A and 1B could, provided proper account is taken of the varying orientation of rooms on each plot, arrange for his heating sub-contractor to prefabricate and install satisfactory heating systems for his houses.

2

Pipe Runs and Radiator Heat Emissions

The next step in design is to plot the pipe runs on the plans and to size the radiators. Pipe runs should be as short and direct as the building layout and structural obstacles permit. The heat to be provided by the radiators will be equal to the calculated heat loss, less the heat emitted from exposed pipework in the rooms.

Table 4 gives the heat emission from single runs of pipe when fixed on brackets close to the walls as in heating practice.

The heat emission from radiator and pipe surfaces depends upon the difference between the mean temperature of the water in the system, and the room temperature. For a gravity circulation operating at the usual temperatures of 82°C flow and 60°C return, the differential upon which a radiator in a room at 16°C is sized is

$$\frac{82 + 60}{2} - 16 = 55°\text{C}$$

For pumped circulations the most suitable temperature drop is in the 11°C to 16°C range. The temperature differential affects the pipe sizing in that it determines the rate of water circulation and therefore the frictional resistance of the circuit. The effect of adopting the higher differential of 16°C is to lower the water circulation rate, thereby reducing friction. This permits longer circuits for equal pipe sizes than is possible with an 11°C temperature drop. A 16°C drop, however, produces a lower difference between the mean water temperature in the system and the room temperature, which means that larger radiators requiring 19 mm connections may be necessary in certain rooms.

Table 4
Heat emission in watts per lineal metre, from uninsulated unpainted copper pipes (BS 659) run singly adjacent to walls as in domestic small-bore central heating, and installed in heated and unheated spaces

Nominal bore of pipe mm	*Space temperature*						
	21°C	18°C	16°C	13°C	10°C	4·4°C	−1°C
(1)	(2)	(3)	(4)	(5)	(6)	(7)	(8)
6	17	20	21	22	23	24	26
10	21	23	24	25	26	29	31
13	26	28	30	32	34	44	47
19	35	38	40	42	44	60	64
25	44	45	48	52	54	74	79
32	52	54	58	62	63	89	94

Notes
(1) Columns 6 to 8 inclusive will generally apply to pipework in unheated outhouses, roof spaces and floor spaces.
(2) For two-pipe runs, multiply by 0·95.
(3) For vertical pipes, multiply by 0·80.
(4) For insulated pipes (80 per cent efficiency) multiply by 0·20.
(5) For bare painted pipes multiply by 1·55.
(6) For Imperial pipe heat emissions see Table 2 Appendix A.

It is partly for this reason and partly because of the rapid heating up from cold that the 11°C differential is generally adopted for small-bore systems. Another advantage of the lower differential is that it ensures higher water temperatures for the radiators on the extreme ends of circuits. In small-bore systems where the circuits can be kept short (as in small houses), it is not difficult to limit the actual temperature drop to 11°C, and the designer should make quite sure that this limiting factor of 11°C is never exceeded. By using the 'Summary of Heat Losses', the proposed circuits can be checked for temperature drop at this early stage.

The velocity of water in small-bore systems should be limited to a maximum of 0·91 m/s. This is mainly to ensure freedom from noise and vibration, and to keep the pump head within reasonable limits. In order to satisfy this limit, the potential heat loadings of small-bore copper pipes must not exceed the amounts in Table 5.

Table 5

Maximum recommended heat-flow and water-flow rates of copper pipes (BS 659) used for small-bore central-heating systems in houses. Water velocity = 0·914 m/s; temperature drop in circuit = 11·0°C

Nominal bore of pipe mm	*Maximum heat flow rate kW*	*Maximum water flow rate kg/s*
6	1·58	0·034
10	3·34	0·072
13	5·73	0·123
19	12·48	0·268
25	21·69	0·466
32	33·30	0·716

For Imperial values see Table 3 Appendix A.

The SI equivalent of the Imperial mass flow rate pounds per hour (lb/h) is the kilogramme per second (kg/s) and

$$1 \text{ lb/h} = 1{\cdot}26 \times 10^{-4} \text{ kg/s}$$

As stated in Chapter 1, the basic metric unit of heat is the kilo-calorie (kcal), the SI heat unit being the joule (J).

If 1 kcal = 4187 J, then
1 kcal/s = 4187 J/s = 4187 watts,
and the SI mass flow rate W will be given by the expression:

$$W = \frac{\text{watts}}{\Delta t \times 4187} \text{ kg/s} \qquad (1)$$

where $\triangle t$ is the temperature difference in degrees C between the flow and return water in the system. The probable temperature drop in any circuit will be given by

$$\triangle t = \frac{\text{watts}}{W \times 4187} \,^{\circ}\text{C} \qquad (2)$$

In Imperial terms the mass flow rate is

$$W = \frac{\text{Btu/h}}{\triangle t}$$

Where W = the mass flow rate in pounds per hour
$\triangle t$ = the temperature difference in degrees F

Example

Find the mass flow rate W for the circuit serving the hall, lounge and bedrooms 2 and 3 in Figs. 2A and 2B.

The system temperature drop is 82°C − 71°C = 11°C, and from the heat-loss summary, the circuit heat load is:

Heat loss		5027
Add 10 per cent for pipes not in rooms		503
Total heat load	=	5530 watts

and the mass flow rate is

$$W = \frac{5530 \text{ watts}}{11 \times 4187} = 0{\cdot}12 \text{ kg/s}$$

The mean water temperature in this small-bore one-pipe heating system will be the generally adopted standard of:

$$\frac{82^{\circ}\text{C} + 71^{\circ}\text{C}}{2} = 76{\cdot}5 \simeq 77^{\circ}\text{C}$$

which is also assumed to be the mean radiator temperature.

The probable temperature drop at a maximum water load of 0·123 kg/s (see Table 5) will be

$$\triangle t = \frac{5530}{0{\cdot}123 \times 4187} = 10{\cdot}74^{\circ}\text{C}$$

The calculated heat load for the lounge circuit at 5530 watts or

5·53 kW is less than the 5·73 kW maximum recommended (Table 5), and is thus within the noise and vibration limit.

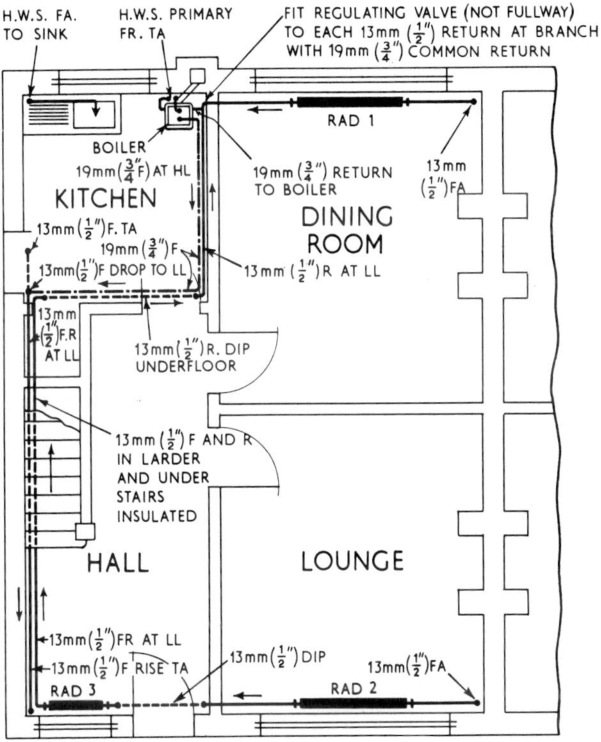

REFERENCE

—·—·—·— PIPE AT HL	FB = FROM BELOW
———— PIPE AT LL	TB = TO BELOW
- - - - - - PIPE UNDER FLOOR	FA = FROM ABOVE
F = FLOW PIPE	TA = TO ABOVE
R = RETURN PIPE	HL = HIGH LEVEL
FR = FLOW AND RETURN	LL = LOW LEVEL

Fig. 2A *One-pipe small-bore heating system: ground floor*

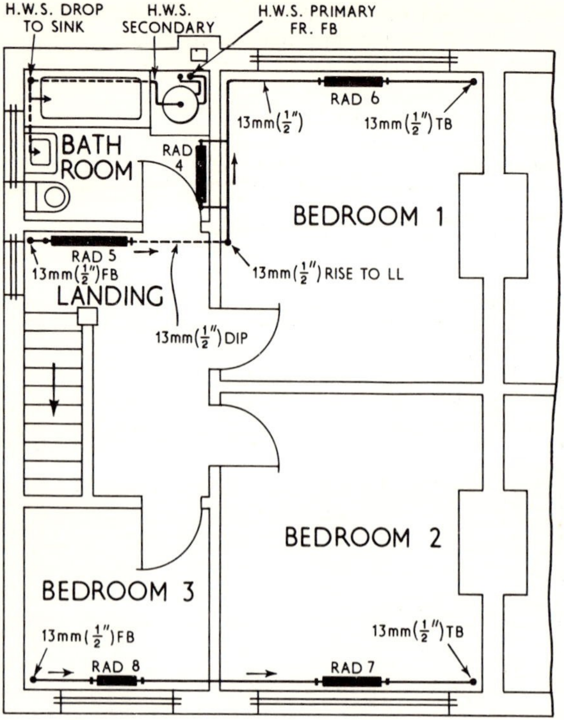

Fig. 2B *One-pipe small-bore-heating system: first floor*

The temperature differences upon which the radiators are to be sized, and the pipe heat emissions calculated, may now be determined.

Dining room and lounge:

$$77°C - 21°C = 56°C \text{ temperature difference}$$

Hall, bathroom, landing:

$$77°C - 16°C = 61°C \text{ temperature difference}$$

Bedrooms:

$$77°C - 13°C = 64°C \text{ temperature difference}$$

Table 6 gives the heat emissions for a typical range of steel radiators for the temperature differences encountered in house warming. Emissions vary slightly according to make, and the manufacturers ratings should be consulted when sizing for a particular make of radiator.

Table 6

Heat emission in watts per square metre from steel radiators when used for domestic small-bore central-heating systems operating at 82°C flow and 71°C return. Radiator mean temperature = 77°C.

Radiator pattern	*Space temperature*			
	21°C	18°C	16°C	13°C
(1)	(2)	(3)	(4)	(5)
Single-panel	605	648	685	726
Double-panel	520	556	587	625
Treble-panel	460	493	520	556
Two-column	500	534	562	600
Three-column	520	556	587	625
Four-column	505	536	568	600
152 mm skirting heaters*	363	395	426	468

**Note:* outputs for skirting heaters are given in watts per metre run. For Imperial values see Table 4 Appendix A.

Before the radiators can be sized, the piping circuits are decided and plotted on the drawings Figs. 2A and 2B. Referring to the plans it is seen that a 19 mm flow pipe leaves the boiler, rises to high level in the kitchen, and splits to two 13 mm flows still at high level. One 13 mm circuit rises from the kitchen to low level on the first-floor landing, serving the landing and bathroom radiators on its way to bedroom 1. After serving the radiator in bedroom 1, the 13 mm pipe drops to feed the dining-room radiator before returning at low level to the kitchen where it is joined by the return from the second 13 mm circuit which serves bedrooms 2 and 3, the lounge, and the hall. A garage built on the south wall could be heated by diverting the 13 mm return on its way to the kitchen. (The design temperature for a private garage is 7°C with one air change per hour.)

Having settled the pipe runs we are in a position to size the radiators, and to make the process understood we will ascertain the preliminary size of the dining-room radiator.

First, carefully measure the pipes in the room. The 13 mm pipe drops from bedroom 1 through the floor thickness and continues at low level at 125–150 mm from the floor, and changes direction to run along the skirting and through the wall into the kitchen. The ceiling is 2·44 m high so that allowing for the thickness of the first floor structure, the vertical drop will be 2·44 m. The pipe falls in the corner of the dining room and, allowing for wall clearance, one bend, and for thickness of the kitchen wall, the length of the horizontal pipe run will be 3·35 m which is the width of the room. The total length of pipe in this room is, therefore, 5·79 m.

Pipes passing through walls and under floors do not emit much heat to the rooms, especially when insulated. In this case, however, and in single-pipe small-house heating generally, the relative length of hidden piping is so small it can safely be credited as adding to the general warmth of the rooms.

All pipes in the house will be painted to match the walls, and for the dining room we must calculate the heat emission from 5·79 m of 13 mm nominal-bore painted copper pipe run near the walls. The temperature difference between the water in the pipes, and the air of the room is 55°C and the heat emission from 13 mm copper pipe at this temperature difference is read from Table 4 as 26 W/m. As already stated the pipe will be painted after fixing and to allow for this the factor of 26 W/m is multiplied by a constant of 1·55.

Note. Table 4 is based on the emissivity of unpainted pipes. When copper pipes are painted their emissivity is increased by approximately 50 per cent.

The probable heat emission from radiator connections may also be taken into account. The emission from the connections of radiators served by single-pipe circuits run at low level in the rooms, can be taken as two thirds of the amounts in Table 7.

Heat emission from pipes in the dining room:

Heat from mains	$= 5{\cdot}79\text{ m} \times 26\text{ W/m} \times 1{\cdot}55 =$	233
Heat from 13 mm radiator connections	$= \dfrac{29\text{ W (Table 7)} \times 2}{3} =$	19
Total heat emission from pipes	$=$	252 W

and the emission for provisional radiator sizing will be:

Dining-room heat loss	=	1781
Less heat emission from pipes	=	252
Calculated radiator heat emission	=	1529 W

Table 7

Average heat emission in watts from radiator connections when the radiator is fixed above and adjacent to concealed horizontal mains. The heat emission from long final sub-circuits should be estimated separately using Table 4.

Nominal bore of pipe mm	*Space temperature*			
	21°C	18°C	16°C	13°C
(1)	(2)	(3)	(4)	(5)
6	20	21	23	24
10	24	26	27	26
13	29	32	34	36
19	39	43	45	47

For Imperial values see Table 5 Appendix A.

A single-panel radiator, which for a temperature difference of 56°C has an emission of 605 W/m², is chosen (Table 6, col. 2).

$$\text{Provisional radiator surface} = \frac{1529\ \text{W}}{605\ \text{W/m}^2} = 2{\cdot}53\ \text{m}^2.$$

The Two-Pipe System

Figs. 3A and 3B show a two-pipe small-bore heating arrangement suitable for the house in Figs. 1A and 1B, the fabric and air-change heat losses being as calculated in Chapter 1. Horizontal mains, branch mains and final sub-circuits to the individual room heaters are run below the floor surface.

Risers to the first floor are enclosed in sealed casings, and in good-class work the casings are filled with insulating material to a thickness of 50 mm around the pipes. All mains, branch mains and final sub-circuits including concealed lengths of radiator connections are insulated to a minimum efficiency of 80 per cent.

Heat emitted from circulating pipes in unheated spaces including outhouses, roof and floor spaces, represents waste, which increases to no good purpose the load on the boiler and circulatory system, inflates both capital and running costs, and generally lowers the efficiency of the installation.

Columns 6 to 8 inclusive of Table 4 may be used when estimating the heat emission from pipes in unheated spaces, column 6 for unventilated floor spaces, column 7 for ventilated floor spaces, and column 8 for roof spaces, outhouses and similar unheated spaces.

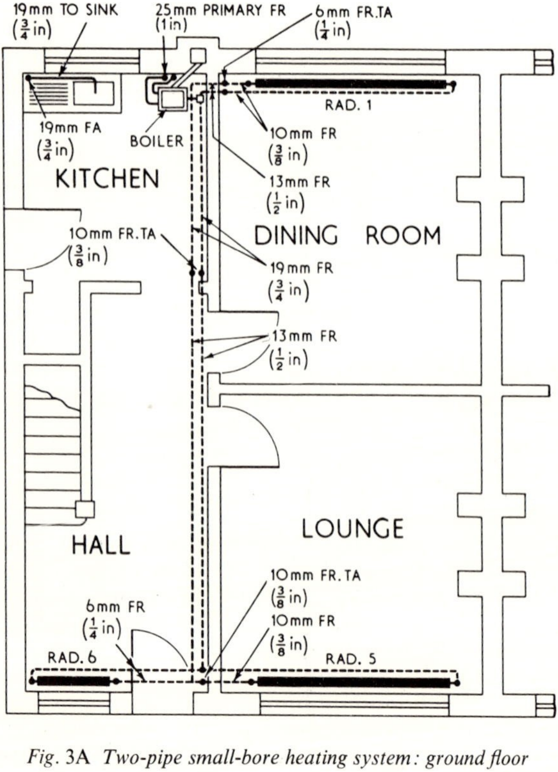

Fig. 3A *Two-pipe small-bore heating system: ground floor*

Radiator emission for a two-pipe system is based on the room heat loss, less the small emission to the room of the exposed portions of the radiator connections. For the small domestic installations which form the subject of this book, the average length of vertical TBOE radiator connections per radiator, exposed in the room when the circulating mains are below the floor surface, is 1 m. Invariably, the mains for both one- and two-pipe installations are run directly beneath the room heater, and where the mains are below the floor surface, the length of concealed radiator connection to allow when pipe sizing, and when estimating pipe heat emission, is 0·6 m vertical and 0·4 m of horizontal pipe per radiator.

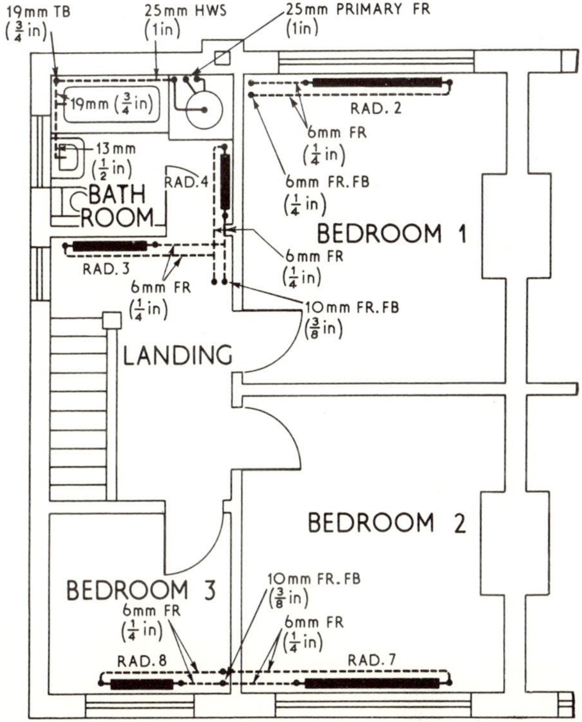

Fig. 3B *Two-pipe small-bore heating system: first floor*

Where, as in the one-pipe system (Figs. 2A and 2B), the mains are run on the wall surface at skirting level and pass under the room heaters, allow an average length per radiator of 0·6 m for vertical TBOE connections.

When BOE connections are used, the length of vertical pipe exposed in the room may be taken as 0·4 m per radiator for both systems, except where the radiator is not adjacent to the mains, in which case the length will be measured from the scheme drawings or on the site.

Table 7 gives heat flow rates from radiator connections; it is computed from Table 4 for the pipe lengths suggested in the foregoing observations and will help in assessing the heat emission from radiator connections. Provisional radiator sizing for a concealed two-pipe system, again using the dining room as our example, is as follows:

Dining-room heat loss	=	1781
Less heat emission from 10 mm TBOE radiator connections (Table 7, col. 2, line 2)	=	24
Calculated radiator heat emission	=	1757 W

$$\text{Provisional radiator surface} = \frac{1757\ \text{W}}{605\ \text{W/m}^2} = 2{\cdot}90\ \text{m}^2$$

3

The Final Sizing and Positioning of Radiators

In Chapter 2 the preliminary sizing* of the dining-room radiator was decided at 2·53m^2. For final sizing the manufacturer's catalogue, which includes information on the actual radiator heating surface according to height and pattern, must be referred to. Before the final choice can be decided from the catalogue, the most suitable radiator height must be settled. This is a most important decision which will depend upon the size of the radiator and its position in the room.

Determining Radiator Height

When the radiator is to be fixed under a window, the sill height will determine the radiator height, and when the radiator must be accommodated on a plain wall surface, the free space available will dictate the height. For example, it may be necessary to place a panel radiator of, say, 1·86 m^2 on a wall which is mainly occupied by furniture restricting the free space to, say, 1·3 m. A 610 mm high panel radiator of 1·86 m^2 heating surface is approximately 1·37 m long, to which add 178 mm for connections and clearance to give an overall length as fixed of 1·55 m and too long for the available wall space. The length of a 760 mm high panel radiator of 1·86 m^2 surface is approximately 1·12 m, plus 178 mm for connections and clearance, to give a fixing length of 1·27 m. A 760 mm high radiator is suitable.

Where free space and aesthetic considerations allow, a low radiator is more suited to house heating, especially where 2·44 m ceilings are encountered. High radiators give a higher temperature

*For the one-pipe system

gradient between floor and ceiling, which is not good in rooms where the occupants spend most of their time sitting at heights of between 300 and 460 mm from the floor. For this reason also, wall radiators should be mounted just above the skirting with approximately 130 mm between the centre line of the bottom tappings and the main.

In houses, where the occupants are more often than not sitting, radiators should if possible be sited under windows. This is especially advisable because a radiator under a window counteracts the cold down draught produced by the cooling effect of the glass, by projecting a rising warm air current across the window. Radiators fixed on walls opposite windows have the opposite effect. In this case warm air currents which are projected from the radiator travel horizontally across the ceiling to the upper window area, where they are suddenly cooled and fall at increased velocity to the lower parts of the room.

Radiators under windows need very careful planning and installation if they are not to constitute an eyesore. Although the heating fitter is often faced with bad building finishes such as unlevel window sills, untrue walls, and uneven plaster work which he cannot always cope with, the heating designer can at least ensure that the radiator he specifies can in fact be fixed below the sill in a sightly manner without protruding some distance above it. Both he and the fitter can also make certain that the radiator is positioned correctly on the vertical centre of the window opening.

Positioning Radiators under Windows

Fig. 4 and Table 8 are included to assist the designer in choosing the correct height of radiator for house installations, by indicating the minimum window-sill height necessary to accommodate properly the standard radiator heights.

Column-type radiator heights as given by the maker's catalogue apply to radiators with feet, the measurements being from the floor on which the radiator stands to the top of the radiator. When column radiators are intended for wall fixing, they are delivered without feet and the actual height is consequently something less than the catalogue dimension. This reduction varies with make of radiator and is usually between 38 and 76 mm, depending upon type and pattern. For practical purposes, in Table 8, a reduction of 76 mm has been allowed; e.g. the height of a 760 mm high column radiator for wall fixing is taken as 684 mm when arriving at the minimum sill height.

Table 8

Minimum window-sill heights for wall-fixed radiators in dwellings

(1)	*Column radiators: catalogue height mm* (2)					*Panel radiators: catalogue height mm* (3)			
	356	457	610	760	914	279	457	610	760
A Top of radiator to top of window sill (mm)	50	50	50	50	50	50	50	50	50
B Bottom of radiator to finished floor (mm)	150	150	150	150	150	150	150	150	150
C Top of window sill to finished floor (minimum height in millimetres)	483	584	737	889	1041	483	660	813	965

Note: this table is to be read in conjunction with Fig. 4.

In the case of panel radiators, the radiator height for wall-fixing purposes is the actual radiator height given in the maker's catalogue, e.g. a 760 mm high panel radiator for wall fixing is taken as being 760 mm high.

In the present example, Figs. 2A and 2B, the radiators are placed under the windows, the only exception being the bathroom where it is not possible to adopt this position. The sill height in all rooms is 840 mm and reference to Table 8, column 9, indicates that 610 mm high panel radiators can be accommodated. Panel radiators 457 mm high would look too low under 840 mm sills and 760 mm radiators would require not less than 965 mm window sills.

Line 1 of Table 8 gives the equivalent metric heights of the radiators to the nearest whole number.

Radiator dimensions vary according to the designs of the several British manufacturers, and the minimum required sill height to accommodate any radiator will be given by one or other of the following expressions:

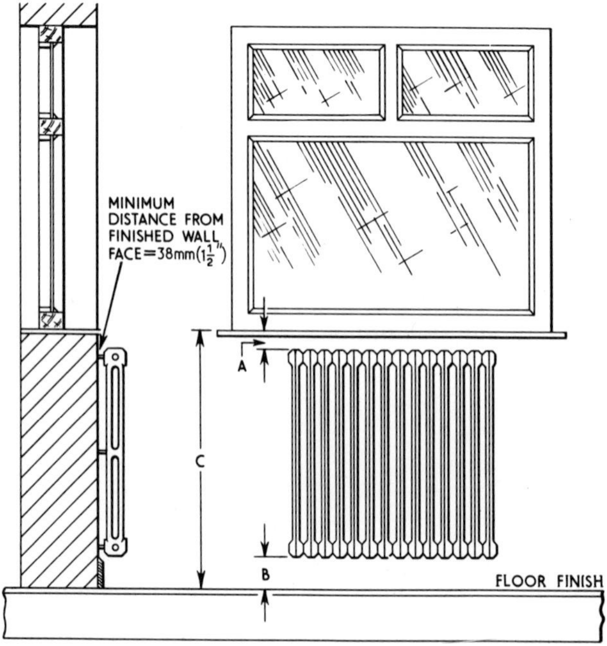

Fig. 4 Radiator fixed under window (see Table 8)

(1) For a wall fixing panel radiator, the sill height *C* (Fig. 4) will be:

$$C = HR + (A + B, \text{Fig. 4})\ \text{mm}$$

Where *HR* is the catalogue height of the radiator in millimetres.

(2) For a wall-fixing column radiator:

$$C = HR + (A + B) - 76 \text{ mm}$$
$$= HR + 200 \text{ mm} - 76$$
$$= HR + 124 \text{ mm}$$

(3) For a radiator on legs:

$$C = HR + 50 \text{ mm}$$

Where *HR* is the height of the radiator from the finished floor surface.

The Final Radiator Sizing

The scheme sheet Fig. 5 summarises the radiator-sizing process for the one-pipe system (Figs. 2A and 2B). Provisional heating surfaces calculated as explained in Chapter 2 are noted in column 6. When, as invariably happens, the provisional size comes between two catalogue sizes, the radiator with the greater heating surface is selected and noted in column 7, which for the dining room is 2·56 m^2.

The final emission from dining-room pipes and radiator will be:

Radiator emission = 2·56 m^2 × 605 W/m^2	=	1549	(col. 10)
Heat emission from pipes in room	=	252	(col. 4)
Total heat emission to the dining room	=	1801 W	(col. 11)

The heat emission from pipes and radiators for all rooms is now determined, and the radiators sized as explained, for the remaining rooms, the results being carefully tabulated in the manner of Fig. 5. This tabulated information is kept in the office job file for reference during the design and installation stages, also for reference should a query arise later concerning the performance of the completed installation.

Radiator Sizing for the Two-Pipe System

A radiator final sub-circuit in a two-pipe system (Figs. 3A, 3B and 10) is simply an extension of the main circuit, and is, therefore, subject to the same circulating pressure.

Because of this, the two-pipe pumped system permits much greater use of 13 mm-bore tube for branch mains and final sub-circuits. Also, the improved design and increased duty range of

(1) *Room name and temperature to nearest whole number °C*	(2) *Temperature difference between heating surface and room °C*	(3) *Room heat loss watts*	(4) *Heat from pipes watts*	(5) *Calculated heat from radiator watts (col. 3 − col. 4)*	(6) *Calculated radiator surface m² divide col. 5 by Table 6*	(7) *Catalogue radiator surface m²*	(8) *Radiator height mm*	(9) *Radiator length m*	(10) *Actual heat from radiator watts (col. 7 × Table 6)*	(11) *Total heat emission to room watts (col. 4 + col. 10)*
Dining room 21	56	1781	252	1529	2·53	2·56	610	1·80	1549	1801
Lounge 21	56	1792	252	1540	2·55	2·56	610	1·80	1549	1801
Hall 16	61	1097	443	654	0·96	1·06	610	0·72	726	1169
Kitchen 16		714								
Bathroom 16	61	630	20	610	0·89	1·06	610	0·72	726	746
Landing 16	61	1025	90	935	1·36	1·37	610	0·96	938	1028
Bedroom (1) 13	64	1371	282	1089	1·50	1·53	610	1·07	1111	1393
Bedroom (2) 13	64	1367	185	1182	1·63	1·71	610	1·19	1242	1427
Bedroom (3) 13	64	771	140	631	0·87	1·06	610	0·71	769	909
									Total = 10·274 kW	

Fig. 5 Domestic heating scheme sheet for one-pipe small-bore system (in SI units).

(1) Room name and temperature °C		(2) Temperature difference °C	(3) Room heat loss watts	(4) Provisional size of radiator connections mm (Table 5)	(5) Heat from radiator connections watts (Table 7)	(6) Calculated heat from radiator watts (col. 3 − col. 5)	(7) Calculated radiator surface m² (Divide col. 6 by Table 6)	(8) Catalogue radiator surface m²	(9) Radiator height mm	(10) Radiator length m	(11) Actual heat from radiator watts (col. 8 × Table 6)	(12) Total heat emission to room watts (col. 5 + col. 11)
Dining room	21	56	1781	10	24	1757	2·90	2·99	457	2·743	1809	1833
Lounge	21	56	1792	10	24	1768	2·92	2·99	457	2·743	1809	1833
Hall	16	61	1097	6	23	1074	1·54	1·54	610	1·080	1055	1078
Kitchen	16		714									
Bathroom	16	61	630	6	23	607	0·87	1·03	610	0·721	706	729
Landing	16	61	1025	6	23	1002	1·45	1·54	610	1·080	1055	1078
Bedroom (1)	13	64	1371	6	24	1347	1·86	1·95	457	1·803	1416	1440
Bedroom (2)	13	64	1367	6	24	1343	1·86	1·95	457	1·803	1416	1440
Bedroom (3)	13	64	771	6	24	747	1·03	1·17	457	1·067	849	873
											Total =	10·304 kW

Fig. 6 Domestic heating scheme sheet for two-pipe small-bore system (in SI units)

small-bore circulating pumps allow a more general use of 6 mm- and 10 mm-bore tube for branch circuits and for final sub-circuits to radiators.

The scheme sheet Fig. 6 tabulates the radiator-sizing process for the two-pipe installation of Figs. 3A and 3B on the basis of building heat losses calculated in Chapter 1.

Reference to Table 5 shows that 6 mm and 10 mm radiator connections cope with the room heat flow rates (col. 3, Fig. 6,) without exceeding the recommended water velocity limit of 0·9144 m/s (3 ft/s).

Heat emission factors for the radiator connections are taken from Table 7 and noted in column 5 of Fig. 6. The calculated radiator surface A for the dining room (Fig. 6) is

$$A = \frac{1781 \text{ (col. 2)} - 24 \text{ (col. 5)}}{605 \text{ (Table 6)}}$$

$$= \frac{1757}{605} = 2{\cdot}90 \text{ m}^2$$

Compared with the calculated radiator surface for the dining room (Fig. 5) which for the one-pipe system is

$$A = \frac{1781 \text{ (col. 3)} - 252 \text{ (col. 4)}}{605 \text{ (Table 6)}}$$

$$= \frac{1529}{605} = 2{\cdot}53 \text{ m}^2$$

This thermal advantage of the one-pipe system over the two-pipe system, which accrues from the exposed 13 mm circulating mains in the rooms and the lower installation costs of the one-pipe method, should not be disregarded.

Where a client is agreeable and the house suitable, a well designed one-pipe system carefully installed by good experienced heating craftsmen, not working against the clock, will give very effective space heating at the lowest cost.

Because the whole system is subject to positive pump circulation which permits a more extensive use of 13 mm-, 10 mm- and 6 mm-bore pipes for branch and final sub-circuits, the two-pipe method has a much wider application for domestic heating than the one-pipe system.

Fig. 6 gives the radiator sizing process, and taking the dining room as our example the detailed sizing is as follows.

Column 1 Room name and temperature: 21°C

Column 2 Temperature difference Δt between the surface of radiator and room

$$t = \frac{82°C + 71°C}{2} - 21°C$$

$$= 76{\cdot}5 - 21 = 55{\cdot}5$$

$$\simeq 56° C$$

Column 3 Room heat loss = 1781 W

Column 4 Provisional size of radiator connections taken from line 2, col. 2 of Table 5, 16 mm-bore pipe has a maximum heat flow rate of 3340 watts (3·34 kW) for a velocity of 0·9144 m/s.

Column 5 Heat emitted to room by radiator connections; from line 2, col. 2 of Table 7, 24 watts.

Column 6 Calculated heat emission from radiator = col. 3 − col. 5 = 1781 − 24 = 1757 watts

Column 7 Calculated radiator heating surface

$$= \frac{1757 \text{ (col. 6)}}{605 \text{ (Table 6)}} = 2{\cdot}90 \text{ m}^2$$

Column 8 Nearest suitable radiator surface from manufacturer's catalogue is 2·99 m².

Columns 9 and 10 The height and length of the radiator must be considered in relation to its suitability for siting under window as per Table 8, or other available wall space. This point is very important and must be settled before the actual catalogue surface can be confirmed.

The more positive circulation provided by the two-pipe system permits the use of short radiators of double- or treble-panel pattern and column radiators where space is limited and the required radiator emission is high, and which would fail to emit their full rated output if dependent upon gravity circulation (see Chapter 5).

A radiator of output not less than the amount in column 6 must be chosen, and, as already stated, when the calculated heating surface (col. 7) comes between two catalogue radiator surfaces, the radiator having the larger surface is chosen.

Column 11 The actual heat emission from the radiator is the product of the surface area A (col. 8) and the heat flow rate in watts per square metre, and substituting in the present case we have

$$A = 2{\cdot}99 \text{ m}^2$$

and from Table 6, the intensity of heat flow rate $= 605$ watts/m^2 then q_r the heat emission from the radiator to the room will be

$$q_r = 2{\cdot}99 \times 605 = 1809 \text{ watts}$$

Column 12 If q_p is the heat emitted by pipes exposed in the room, in this case the radiator connections (since all other circulating pipes are concealed), the total heat emission Q to the room is

$$\begin{aligned} Q &= q_r + q_p \\ &= 1809 + 24 \\ &= 1833 \text{ watts} \end{aligned}$$

4

Pipe and Pump Sizing

Fig. 7 is a diagrammatic arrangement of the small-bore one-pipe heating system for the ground- and first-floor plans shown in Figs. 2A and 2B. It is representative of small-bore systems suitable for warming small dwellings of up to 93m² floor area (1000 ft²).

Two 13 mm sub-circuits are usually necessary, these being fed as shown in Figs. 2A, 2B and 7, by a main 19 mm circuit from the

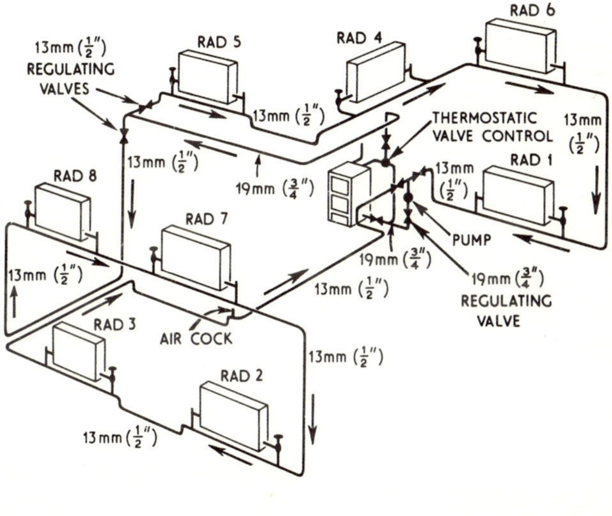

Fig. 7 Diagrammatic arrangement of one-pipe small-bore system

boiler. As previously stated, a 19 mm main circuit is suitable for jobs up to 11·72kW, which is adequate for the type of house under consideration, and which will provide the greater proportion of future house-heating business.

The Pump

The successful application of small-bore pipes for centrally heating domestic premises is entirely due to the development of a small glandless, electrically-driven centrifugal pump, designed to circulate small quantities of water against relatively high heads.

The function of the pump is to circulate water to replace the heat emission from the pipes and radiators, against the resistance offered by pipes, pipe fittings, valves, the temperature control valve, the pump outlet and the boiler. A list of pipes and resistances should be carefully prepared. The pipes are measured from the plans (Figs. 2A and 2B), and resistances which include all valves and other pipe fittings can be 'taken off' and listed from a diagram of the proposed installation (Fig. 7).

Boiler and Pump Connections

Boiler and pump pipe connections contribute considerably to the friction loss of the system, and an isometric diagram of the proposed connections will greatly assist the designer in enumerating these resistances. Fig. 8 shows the working arrangement of the boiler and pump connections for the installation in Figs. 2A and 2B, and the resistances can be taken off this sketch and listed. Table 9 includes resistance values for valves and copper pipe fittings and covers the sizes of pipework used for small-bore house heating.

The pipe-sizing chart Fig. 9 gives the frictional head loss in newtons per square metre per metre run of copper pipe at velocities up to a maximum of 0·91 m/s, and embraces the pipe sizes most commonly used for small-bore heating, namely 6 mm, 10 mm, 13 mm, 19 mm, 25 mm, and also 32 mm, which is sometimes installed for primary hot-water circulations.

The lengths of 13 mm pipe as measured from the plans, Figs. 2A and 2B, are as follows:

Dining-room circuit	= 15·24 m
Lounge circuit	= 34·75 m

Length of 19 mm flow and return pipes in kitchen, including boiler and pump connections = 6·40 m

Referring to Figs. 7 and 8, and Table 9, complete the schedules of resistances and record the equivalent pipe lengths.

Capillary fittings will be used for this installation.

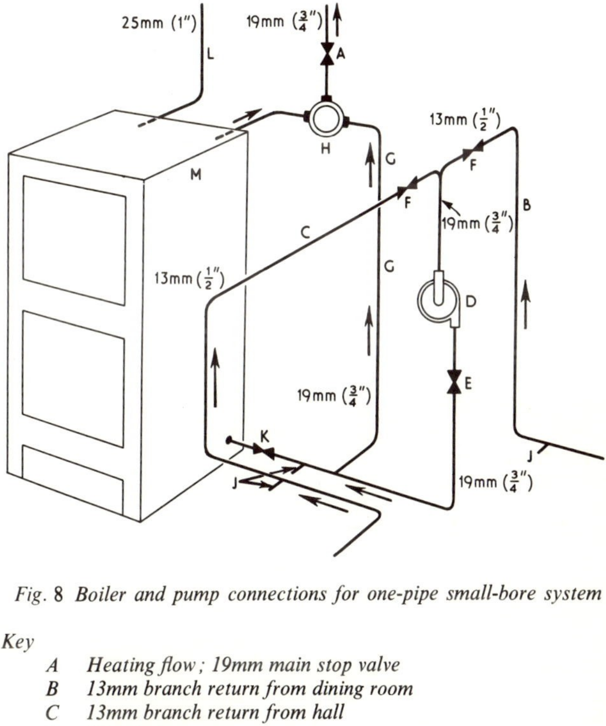

Fig. 8 Boiler and pump connections for one-pipe small-bore system

Key

- *A* *Heating flow; 19mm main stop valve*
- *B* *13mm branch return from dining room*
- *C* *13mm branch return from hall*
- *D* *Circulating pump*
- *E* *Pump regulating valve on pump discharge*
- *F* *Return branch stop valves*
- *G* *Return bypass to mixing valve*
- *H* *Automatic mixing valve*
- *J* *Tee branch for empty cock*
- *K* *Heating return 19mm main stop valve*
- *L* *Primary flow pipe to HWS indirect cylinder*
- *M* *Boiler*

Table 9

Resistance of copper fittings etc. normally used for small-bore pumped hot-water heating systems: Equivalent length of straight pipe in metres; comp. = compression fitting; cap. = capillary fitting.

Description of local resistances	*Nominal pipe bore (mm)*					
	6	10	13	19	25	32
Regulating valve fully open			7·32	8·53	12·20	
,, ,, ¾ ,,			12·20	18·29	21·34	
,, ,, ½ ,,			27·43	45·72	54·86	
,, ,, ¼ ,,			152	290	305	
Gate valve (full way)	0·11	0·19	0·25	0·43	0·61	0·79
Globe valve or tap	2·04	3·35	4·48	7·32	10·21	13·11
Radiator valve (angle)	1·03	1·71	2·29	3·81	5·49	7·01
Elbow comp.	0·25	0·43	0·55	0·88	1·22	1·49
Elbow cap.	0·16	0·26	0·37	0·58	0·82	1·07
Bend comp. or cap.	0·12	0·19	0·25	0·40	0·58	0·73
Through tee comp. or cap.	0·21	0·04	0·05	0·08	0·12	0·15
Tee: comp. or cap.						
Flow into branch	0·27	0·46	0·61	0·98	1·37	1·80
Flow out of branch	0·30	0·52	0·67	1·13	1·62	2·07
Reducer tee branch comp. or cap.	0·34	0·55	0·73	1·23	1·74	2·23
Pitcher tee comp.						
Flow into branch	0·18	0·29	0·40	0·67	0·98	1·28
Flow out of branch	0·20	0·30	0·43	0·76	0·98	1·22
Pitcher tee cap.						
Flow into branch	0·22	0·37	0·46	0·73	1·04	1·34
Flow out of branch	0·22	0·37	0·49	0·76	1·40	1·19
Twin elbow comp.						
Flow into branch	1·34	0·22	0·49	0·56	0·70	0·88
Flow out of branch	0·12	0·20	0·44	0·56	0·64	0·83
Twin elbow cap.						
Flow into branch	0·19	0·30	0·40	0·61	0·83	1·01
Flow out of branch	0·19	0·30	0·43	0·67	0·94	1·22
Return bend (long sweep) comp. or cap.	0·15	0·24	0·34	0·50	0·67	0·83
Reducing coupling comp. or cap.	0·06	0·10	0·14	0·23	0·30	0·43
Outlet from pump, tank or cylinder			0·67	1·13	1·62	2·29
Hot-water boiler			2·13	2·74	3·66	4·88
Radiator	0·91	1·52	2·44	3·96	5·49	—

As explained in Chapter 2, the SI mass flow rate W for a hot-water circulation is given by

$$W = \frac{\text{Heat flow rate in watts}}{\Delta t \times 4187} \text{ kg/s}$$

The value of W for each branch circuit will be

Dining-room circuit Comprising, dining room, bedroom 1, bathroom and landing.

$$W = \frac{4968 \text{ (from col. 11, Fig. 5)}}{11 \times 4187}$$

$$= 0{\cdot}108 \text{ kg/s.}$$

Lounge circuit Comprises, lounge, hall and bedrooms 2 and 3.

$$W = \frac{5306 \text{ (from col. 11, Fig. 5)}}{11 \times 4187}$$

$$= 0{\cdot}115 \text{ kg/s.}$$

Heat Emission from Mains

The water circulation required to make good the heat emission from all 13 mm pipes and radiators has now been calculated. There remains the emission from the 13 mm and 19 mm pipes in the kitchen, and also the heat emitted by the external surfaces of the boiler. Water must, of course, be circulated to account for this emission. In the present example the heating pipes in the kitchen will not be insulated, in order to keep this room at a reasonable temperature.

The boiler is fitted with lined insulating jacket and will emit approximately 293W which, added to that emitted by the pipes (724W), gives a total of 1017W. Where an unlined jacket is fitted, the heat loss from the type of small boiler usually installed for small house-warming systems can be taken at about 880W.

The heat loss from the kitchen for a room temperature of 16°C with two air changes per hour will be 714W. Uninsulated high-level heating pipes in the kitchen will also prevent condensation on the ceiling. Primary hot-water pipework in the kitchen should in this case be insulated, both to save fuel and to limit the room temperature during the summer.

Circulating water load for kitchen

$$\frac{1017}{11 \times 4187} = 0{\cdot}022 \text{ kg/s}$$

The Index Circuit

The pump is sized to circulate the water against the resistance of the index circuit, which is that circuit offering the greatest resistance to the flow of water.

In the present example there are two 13 mm circuits fed from a 19 mm main circuit. These are the lounge circuit, 48·48 m plus length of boiler connections, and 19 mm common pipes of 22·46 m equivalent length, making a total equivalent length of 71 m, and the dining room circuit, 27·21 m plus 22·46 m for boiler connections and 19 mm common pipes, total equivalent length

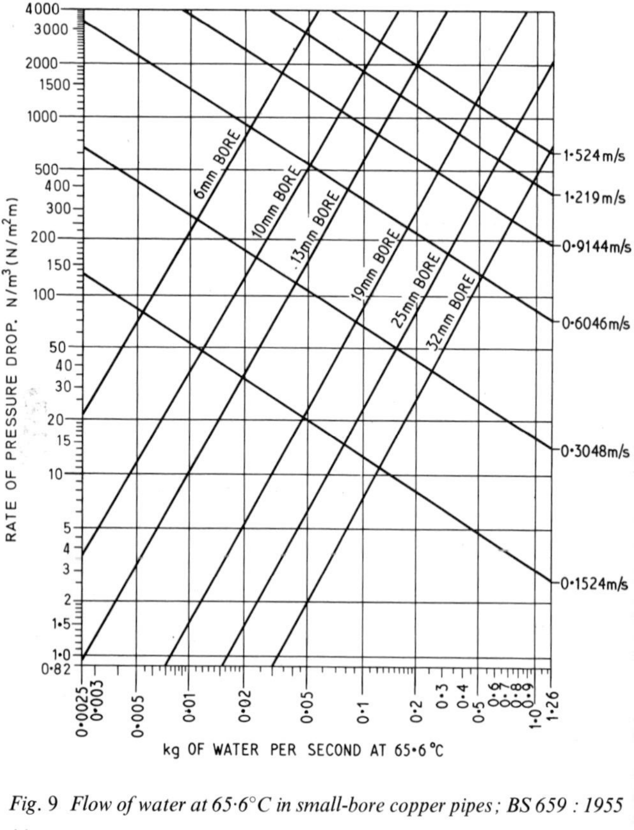

Fig. 9 Flow of water at 65·6°C in small-bore copper pipes; BS 659 : 1955

49·67 m. The lounge circuit is the index circuit and the pump should be sized to circulate the total water load of the system against the resistance of this circuit. The equivalent length of each circuit is calculated in the schedules which follow.

Dining-room sub-circuit: Schedule of effective length

Description of resistance	*No. off*	*Factor Table 9 m*	*Equivalent pipe length, m col. 2 × col. 3*
13 mm regulating valve	1	7·32	7·32
13 mm gate valve	1	0·25	0·25
13 mm bends	7	0·25	1·75
13 mm elbows (forming dips in floor)	4	0·37	1·48
13 mm through tees	10	0·05	0·50
13 mm tee	1	0·67	0·67
Equivalent length due to resistances			= 11·97 m
Measured length of pipe			= 15·24 m
Total effective length of the sub-circuit			= 27·21 m

Boiler connections and pipework in kitchen: Schedule of effective length

Description of resistance	*No. off*	*Factor Table 9 m*	*Equivalent pipe length, m col. 2 × col. 3*
Boiler	1	2·74	2·74
19 mm gate valve	2	0·43	0·86
19 mm regulating valve	1	8·53	8·53
19 mm bends	5	0·40	2·00
19 mm twin elbows	1	0·56	0·56
19 mm through tee	3	0·08	0·24
19 mm pump outlet	1	1·13	1·13
Equivalent length due to resistances			= 16·06 m
Measured length of 19 mm pipe			= 6·40 m
Total effective length of main circuit, boiler and pump connections			= 22·46 m

Lounge sub-circuit: Schedule of effective length

Description of resistances	*No. off*	*Factor Table 9 m*	*Equivalent pipe length, m col. 2 × col. 3*
13 mm regulating valve	1	7·32	7·32
13 mm gate valve	1	0·25	0·25
13 mm bends	11	0·25	2·75
13 mm elbows (forming dips in floor)	8	0·37	2·96
13 mm through tees	9	0·05	0·45
Equivalent length due to resistances		=	13·73 m
Measured length of 13 mm pipe		=	34·75 m
Total effective length of sub-circuit		=	48·48 m

Pressure and Pressure Drop

For pipe and pump sizing, pressure in a water system may be measured in Imperial terms of feet or inches of water (ft H_2O or in. H_2O).

In low-pressure hot-water heating, the pressures are very low and are stated in terms of in. H_2O.

The SI metric equivalent units are the bar and the millibar (mb), 1 bar being equal to 1000 mb. The standard metric equivalent of the in. H_2O is the mm H_2O, and as 1 in. = 25·4 mm, then, in. $H_2O \times 25{\cdot}4$ = mm H_2O.

The SI unit of force is the 'newton' (N), and the unit of pressure is the 'newton per square metre' (N/m^2).

$$1 \text{ mb} = 100 \text{ N/m}^2\text{, and}$$

$$1 \text{ mm H}_2\text{O} = 9{\cdot}807 \text{ N/m}^2\text{, then}$$

$$1 \text{ in. H}_2\text{O} = 25{\cdot}4 \times 9{\cdot}807 = 249{\cdot}1 \text{ N/m}^2$$

In Imperial terms, pressure drop ($\triangle P$) in a pipe circuit is expressed as in. H_2O per foot run, the SI equivalent being N/m^3 (N/m^2m).

We know that 1 in. $H_2O = 249{\cdot}1\ N/m^2$ and also that 1 metre $= 3{\cdot}2808$ ft, from which we deduce that

$$1 \text{ in } H_2O/\text{ft} = 249{\cdot}1 \times 3{\cdot}2808$$
$$= 817{\cdot}6\ N/m^3$$

Estimating the Pump Duty

The duty of the pump in a small-bore central-heating system is to circulate the water load to satisfy the heat emission of the whole installation against the resistance of the index circuit, which is that circuit having the greatest pressure drop or resistance to flow, in this case the lounge sub-circuit plus the boiler and pump connections.

Referring to Fig. 9 the resistance of the index circuit is calculated in the following manner:

Lounge sub-circuit: 48·48 m of 13 mm pipe circulating 0·115 kg of water per second, (0·115 kg/s).

From the 0·115 kg/s point on the 'rate of flow' scale of Fig. 9, project a vertical line to the 13 mm-pipe curve, and from the point of intersection project a horizontal line to the 'rate of pressure drop' scale, and read off the pressure drop in newtons per square metre per metre run of pipe as 690.

The total pressure drop in the lounge sub-circuit is

$$690\ N/m^3 \times 48{\cdot}48\ m = 33451\ N/m^2$$

The main 19 mm flow and return pipe circulates the water load for the whole system, which totals:

13 mm lounge sub-circuit		0·115
13 mm dining-room sub-circuit		0·108
19 mm main circulating pipes, boiler and boiler connections		0·022
Total water load	=	0·245 kg/s

The effective length of the boiler connections and the 19 mm main circulating pipes is 22·46 m, and we have 22·46 m of 19 mm-bore pipe circulating 0·245 kg/s, and from Fig. 9 the pressure drop per metre run is 420 N/m^3, the total pressure loss in the mains being

$$420\ N/m^3 \times 22{\cdot}46\ m = 9433\ N/m^2$$

which added to the resistance of the lounge sub-circuit gives a total resistance for the index circuit of

$$33451\ N/m^2 + 9433\ N/m^2 = 42884\ N/m^2$$

To obtain the final pump duty, the resistance of the temperature-control valve must be allowed for.

Depending on their make and design, the head-loss characteristic of these valves varies under identical operating conditions, and for this reason it is not possible to state a common resistance value in terms of equivalent pipe length, as for standard valves and fittings.

Reputable makers of this type of control are able to furnish the scheme designer with accurate head-loss data, and for the scheme under consideration the valve manufacturer gave the pressure drop across the valve as 1200 N/m^2 at 0·245 kg/s. The total pressure drop in the system may now be estimated at

$$42884\ N/m^2 + 1200\ N/m^2 = 44084\ N/m^2$$

The pump will be required to circulate 0·245 kg of water per second against a total head of 0·44 bar.

In the Imperial system, pump duty is stated in gallons per minute (gal/min) against pressure or head in feet (ft H_2O), and in SI units the duty is expressed in litres per second (litre/s) against pressure in bars or decimal fractions of the bar.

Let W represent the mass flow rate for the whole heating system, and $\triangle P$ the pressure drop in the index circuit, then, in Imperial units, where W is in pounds per hour and $\triangle P$ in in. H_2O,

$$\text{pump duty} = \frac{W}{10 \times 60}\ \text{gal/min at}\ \frac{\triangle P}{12}\ \text{ft}\ H_2O$$

The litre is the volume occupied by a mass of 1 kilogramme of pure water at approximately 4°C (39·2°F), and subject to atmospheric pressure.

In SI units when W is in kilogrammes per second and $\triangle P$ in newtons per square metre,

$$\text{pump duty} = W \times 1\ \text{litre/s at}\ \frac{\triangle P}{10^5}\ \text{bar,}$$

and the pump duty for the one-pipe small-bore installation (Figs. 2A, 2B, 6 and 7), where $W = 0{\cdot}245$ kg/s and $\Delta P = 44084$ N/m² is

$$\text{pump duty} = 0{\cdot}245 \text{ litre/s at } \frac{44084}{10^5} \text{ N/m}^2$$

$$= 0{\cdot}245 \text{ litre/s at } 0{\cdot}44 \text{ bar}$$

The Two-Pipe Small-Bore System

The isometric diagram (Fig. 10) represents the two-pipe installation in Figs. 3A and 3B.

As shown, 19 mm F and R mains are taken off the boiler to serve 13 mm, and 10 mm F and R branch mains feeding the final sub-circuits to individual radiators. For pipe sizing, the system is divided into convenient sections, e.g. section AB rises from section CD to feed radiators 7 and 8.

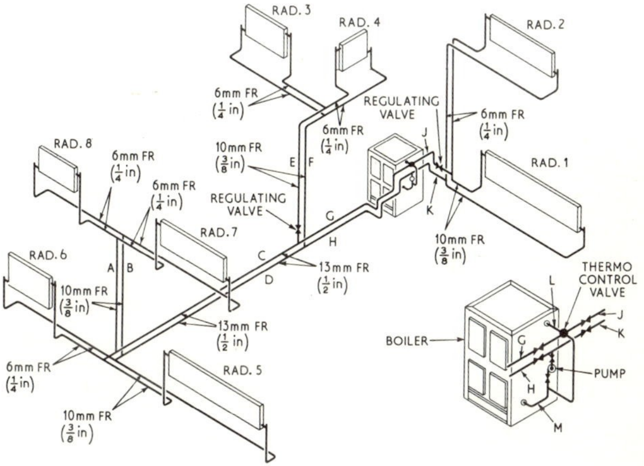

Fig. 10 *Diagrammatic arrangement of two-pipe small-bore system*

Boiler and pump connections

G = 19mm (¾ in)
H = 19mm (¾ in)
J = 13mm (½ in)
K = 13mm (½ in)
L = 19mm (¾ in)
M = 19mm (¾ in)

An orderly and rational method of tabular calculation should be used for pipe and pump sizing, especially when designing two-pipe systems. Fig. 11 illustrates the provisional pipe sizing for the two-pipe installation shown in Figs. 3A, 3B and 10.

(1) *Circuit section reference*	(2) *Heat emission to rooms watts*	(3) *Add 10% for mains watts*	(4) *Heat flow rate watts (col. 2 + col. 3)*	(5) *Provisional pipe size from Table 5 mm*
Radiator 7	1440	144	1584	6
Radiator 8	873	87	960	6
Section AB =	(rad. 7 + rad. 8)		2544	10
Radiator 5	1833	183	2016	10
Radiator 6	1078	108	1186	6
Section CD =	(section AB + rads. 5 and 6)		5746	13
Radiator 3	1078	108	1186	6
Radiator 4	729	73	802	6
Section EF =	(rad. 3 + rad. 4)		1988	10
Section GH =	(section CD + section EF)		7734	19
Radiator 1	1833	183	2016	10
Radiator 2	1440	144	1584	6
Section JK =	(rad. 1 + rad. 2)		3600	13
Section LM =	(section GH + section JK)		11334	19

Fig. 11 Domestic heating scheme sheet: provisional pipe sizing for two-pipe small-bore system (in SI units)

Pipe section and radiator references are noted in column 1, and the room heat emissions (from col. 12, Fig. 6) are entered in column 2. For provisional sizing two-pipe systems having concealed mains and branch mains, add 10 per cent to column 2, enter in column 3. The sum of columns 2 and 3 gives the provisional heat flow rate in column 4. Using the values in column 4, provisional pipe sizes in column 5 are obtained from Table 5 and provide the basis for a more accurate estimate of mains heat emission.

With the aid of Table 4, estimate the heat emission from the insulated pipes in floor spaces and riser casings, the boiler outer surface, and boiler connections, as follows.

Horizontal Pipes
Floor space temperature = 4·4°C.

4·88 m of 19 mm pipe × 60 W/m × 2 × ·95	=	56
10 m of 13 mm pipe × 44 W/m × 2 × ·95	=	84
10·67 m of 10 mm pipe × 29 W/m × 2 × ·95	=	59
Heat emission from horizontal pipes	=	199 W

Vertical Pipes

10·97 m of 10 mm pipe × 29 W/m × 2 × ·95 × ·8	=	48
7·32 m of 6 mm pipe × 24 W/m × 2 × ·95 × ·8	=	28
Heat emission from vertical pipes	=	76 W

Heat emission from boiler and connections = 483 W

Total emission from boiler and mains = 199 + 76 + 483 = 758 W, which is approximately 8 per cent of the room heat emission.

The heat flow rates are now amended as shown in Fig. 12, and the final pipe sizes determined (column 6).

(1) *Circuit section reference*	(2) *Heat emission to rooms watts*	(3) *Add 8% for mains watts*	(4) *Heat flow rate (col. 2 + col. 3) watts*	(5) *Water flow rate* $W = \frac{\text{col. 4}}{11{\cdot}1 \times 4187}$ *kg/s*	(6) *Final pipe size mm*
Radiator 7	1440	115	1555	0·034	6
Radiator 8	873	73	946	0·021	6
Section AB	= (rad. 7 + rad. 8)		2501	0·054	10
Radiator 5	1833	147	1980	0·043	10
Radiator 6	1078	86	1164	0·025	6
Section CD	= (section AB + rads. 5 and 6)		5645	0·123	13
Radiator 3	1078	86	1164	0·025	6
Radiator 4	729	58	787	0·017	6
Section EF	= (rad. 3 + rad. 4)		1951	0·042	10
Section GH	= (section CD + section EF)		7596	0·165	19
Radiator 1	1833	147	1980	0·043	10
Radiator 2	1440	115	1555	0·034	6
Section JK	= (rad. 1 + rad. 2)		3535	0·077	13
Section LM	= (section JK + section GH)		11131	0·242	19

Fig. 12 Domestic heating scheme sheet: final pipe sizing for two-pipe small-bore system (in SI units)

Sizing the Circulating Pump

Using the isometric diagram Fig. 10, take off the number, size and type of fittings (local resistances) and note these in columns 2–11 of Fig. 13 which is a tabular method of calculating the effective length of the system. The items in columns 2–11 are multiplied by the lengths given in Table 9 to produce the 'equivalent length for local resistances' in column 12. The measured length of pipework is obtained from Figs. 3A, 3B and 10, and is entered in column 13 and added to column 12 to give the effective length in column 14.

The pump must be sized to circulate the total water load of the system against the resistance of the index circuit which, as explained earlier, is that circuit offering the greatest resistance to flow.

This installation comprises three main branch circuits:

circuit 1 serving radiators 1 and 2 (the index radiator is no 2).
circuit 2 serving radiators 3 and 4 (the index radiator is no. 3).
circuit 3 serving radiators 5, 6, 7 and 8 (the index radiator is no. 7).

Fig. 14 sets out the pump-sizing process for a two-pipe system. The respective circuits are identified in column 1, and the water flow rates and final pipe sizes, taken from columns 5 and 6 of Fig. 12 are entered in columns 2 and 3 of Fig. 14.

Column 4, the rate of pressure drop at the water flow rate (column 3), is read from Fig. 9. The effective length (column 5) is taken from column 14 of Fig. 13.

Fig. 9 gives the rate of pressure drop in newtons per cubic metre, and the pressure drop in each section is the product of columns 4 and 5, the total pressure drop (or resistance) being progressively recorded in column 7. Comparing the totals shows circuit 3 to have the highest resistance at 35069 N/m^2mb, and this is therefore the index circuit for pump sizing.

From column 5, Fig. 12 the total water flow rate for the system is 0·242 kg/s which for pump-sizing purposes is equal to 0·242 litre/s, and the duty of the pump will be to circulate 0·242 litre/s against a head of 35069 N/m^2 mb (column 7 of Fig. 14).

In round figures the specified duty of the pump is:

$$0{\cdot}24 \text{ litre/s at } \frac{35100}{10^5}$$

$$= 0{\cdot}24 \text{ litre/s at } 0{\cdot}351 \text{ bar}$$

(1) *Pipe reference and size mm*		(2) *Boiler*	(3) *Pump*	(4) *Stop valve fullway*	(5) *Tee branch (out)*	(6) *Tee branch (in)*	(7) *Through tee*	(8) *Bend*	(9) *Elbow*	(10) *Radiator*	(11) *Radiator valve*	(12) *Equivalent length m*	(13) *Measured length m*	(14) *Effective length m*
A	10				1	1						0·98	2·59	3·57
B	10				1	1						0·98	2·59	3·57
C	13					1	2					0·71	5·03	5·74
D	13				1		2					0·77	5·03	5·80
E	10				1				1			0·78	3·05	3·83
F	10					1			1			0·72	3·05	3·77
G	19			1			1	3	2			2·87	3·35	6·22
H	19			1			1	3	2			2·87	3·35	6·22
J	13					1		1	1			1·41	1·52	2·93
K	13				1			1	1			1·47	1·52	2·99
L and M	19	1	1	2	2			2				7·79	1·68	9·47
Radiators														
1	10						2		6	1	2	6·58	4·72	11·30
2	6						2		6	1	2	4·35	9·45	13·80
3	6				1	1			4	1	2	4·18	4·12	8·30
4	6						2		4	1	2	4·03	2·59	6·62
5	10						2		4	1	2	6·06	5·49	11·50
6	6						2		4	1	2	4·03	4·72	8·75
7	6						2		4	1	2	4·03	5·18	9·21
8	6						2		4	1	2	4·03	3·35	7·38

Fig. 13 *Domestic heating scheme sheet: effective length of two-pipe small-bore circuits (in SI units)*

(1) *Pipe reference*	(2) *Pipe size mm*	(3) *Water flow rate kg/s*	(4) *Rate of pressure drop (from Fig. 9) N/m³*	(5) *Effective length m*	(6) *Pressure drop N/m² (col. 4 × col. 5) to nearest whole number*	(7) *Total pressure drop N/m²*
Circuit 1						
Radiator 2	6	0·034	1600	13·80	22080	22080
J	13	0·077	350	2·93	1026	23106
K	13	0·077	350	2·99	1047	24153
Boiler and pump connections L and M	19	0·220	345	9·47	3267	27420
Thermostatic control valve					1400	28820
Circuit 2						
Radiator 3	6	0·025	1080	8·30	8964	8964
E	10	0·042	390	3·83	1494	10458
F	10	0·042	390	3·77	1470	11928
G	19	0·165	210	6·22	1306	13234
H	19	0·165	210	6·22	1306	14540
L and M and thermostatic control valve					4667	19207
*Circuit 3**						
Radiator 7	6	0·034	1600	9·21	14736	14736
A	10	0·054	600	3·57	2142	16878
B	10	0·054	600	3·57	2142	19020
C	13	0·123	760	5·74	4362	23382
D	13	0·123	760	5·80	4408	27790
G					1306	29096
H					1306	30402
L and M and thermostatic control valve					4667	35069

Fig. 14 *Domestic heating scheme sheet: pump sizing for two-pipe small-bore system (in SI units)*
**Index circuit: allow a pressure drop across the valve of 1400 N/m²*

The Mini-Bore System

The mini-bore low-pressure hot-water heating system makes greater use of 6mm ($\frac{1}{8}$ in), 10 mm ($\frac{3}{8}$ in) and 13 mm ($\frac{1}{2}$ in) bore pipes for final sub-circuits, branch circuits and main branch circuits respectively. It is based upon a two-pipe method of distribution from, depending upon load, 19 mm ($\frac{3}{4}$ in) or 25 mm (1 in) manifold pipes, which may be centrally disposed at each floor level, and to which the branch circuits or the final sub-circuits serving the radiators are connected.

Heat load calculation, radiator sizing, pipe and pump sizing exactly follows the procedures already explained in this and preceding chapters for the standard small-bore system.

The basic condition permitting the extended use of the smaller pipes when compared with the standard small-bore circuit is the increase in design water velocity from the recommended 0·9144 m/s, (3 ft/s) to 1·219 m/s (4 ft/s), which enables pipe sizing to be based on the following flow rates, using BS 659, copper pipe:

Pipe bore	*Maximum water flow*	*Heat flow*
6 mm	0·045 kg/s	2·07 kW
10 mm	0·096 kg/s	4·42 kW
13 mm	0·160 kg/s	7·37 kW
19 mm	0·370 kg/s	17·04 kW
25 mm	0·650 kg/s	29·94 kW

The heat-flow rates quoted are related to a temperature differential of 11°C.

Pump Margin

If, as described in the foregoing, the pump has been carefully sized and proper account taken of all resistances and water loadings, a margin is entirely unnecessary. Bad pump sizing involving the addition of a mystical margin of anything from 15 to 25 per cent is the greatest single factor contributing to noise and vibration in small-bore heating installations. It also adds to the risk of counter-flow in the primary hot-water circulation.

Pumps fitted with an integral regulating valve are now on the market, and where a pump not so equipped is to be installed, a specially-designed regulating valve (not an ordinary gate or globe valve) should be fitted to the pump discharge. A regulating valve so fitted will allow the use of a pump having a standard duty somewhat higher than the exact calculated duty. It must be accepted that the manufacturer cannot supply an individual pump to match exactly the head of every small-bore installation. This being so, the need for regulation on the great majority of jobs is plainly apparent.

Fig. 15 is the duty chart for a small-bore pump fitted with a variable-output device by which the pump can be adjusted within its capacity limits. For the average small-bore system, pumps developing heads of 0·18–0·36 bars are suitable. This pressure limit complicates design of the larger installations in that it restricts the length of the 13 mm sub-circuits. To overcome this restriction pumps are being developed with duties of 0·5–0·6 litre/s against heads of up to 0·5 bars.

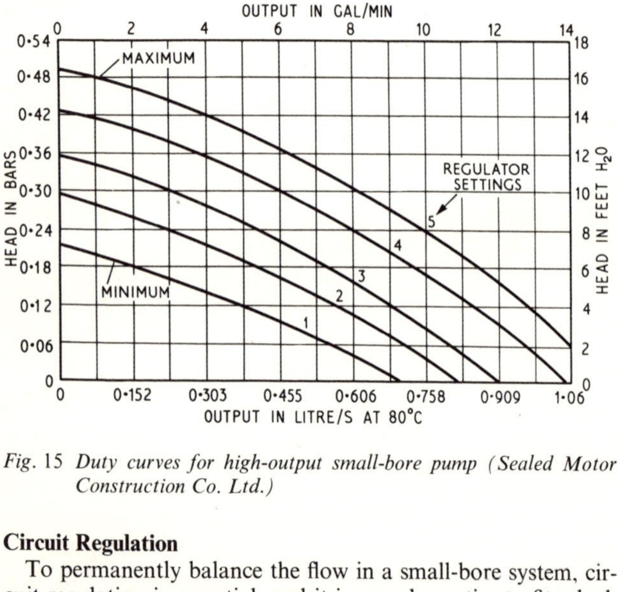

Fig. 15 Duty curves for high-output small-bore pump (Sealed Motor Construction Co. Ltd.)

Circuit Regulation

To permanently balance the flow in a small-bore system, circuit regulation is essential, and it is sound practice to fit a lock shield regulating valve to the flow of every sub-circuit, except the index sub-circuit. Many competent installers fit a regulating valve, (see Figs. 7 and 10) to all sub-circuits, and to the pump outlet, where this is not of the variable-head type.

The two-pipe system Fig. 10, is to be provided with a variable-head pump, and regulating valves should be fitted as shown to pipe *J* of sub-circuit 1 and pipe *E* of sub-circuit 2. A standard fullway stop valve should be fitted to the inlet and outlet of every circulating pump, to facilitate maintenance and eventual replacement. Radiators 5, 6, 7 and 8 can be balanced by adjusting the lock shield valves on the radiator outlets.

In both one-pipe and two-pipe installations, regulation of water flow to individual room heaters by means of a lock shield, or other type of 'tamper-proof' valve fitted to the heater outlet is essential if the system is to give effective service. A wheel valve fitted to the inlet is also necessary to enable the householder to shut off the radiator when he wishes to do so. Furthermore, the

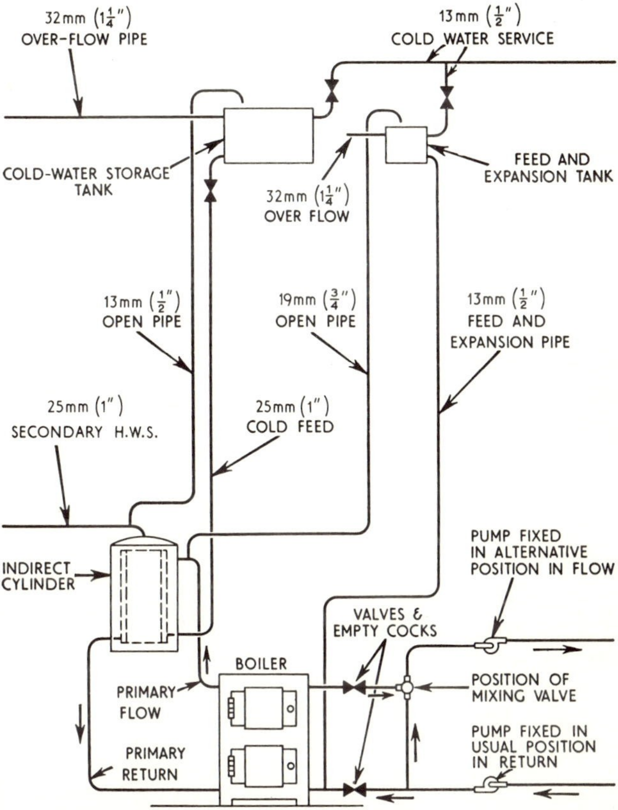

Fig. 16 Central plant for typical combined small-bore hot-water heating and hot-water supply system; note the alternative pump positions

valves fitted to a radiator or other heater should be union pattern, to facilitate removal of the heater for repair, or replacement, or for room decoration.

Position of the Pump

Correct positioning of the pump is most important, and will be dictated by the piping layout. In Fig. 8 the pump is shown fitted into the return pipe. This position is suitable for all installations where the highest circuits are run at low level on the first floor of two-storey houses, as in the present example. Where pipes are carried to high level in top storey rooms, and possibly in the roof space as is sometimes necessary in bungalows, there is a danger when the pump is fitted in the return (as Fig. 8) that the pressure in the high level pipe will be a negative one. When pipes are run in the high level positions mentioned, the pump should be fitted in the flow pipe as shown in Fig. 16. Generally speaking, the flow-pipe position is to be preferred, with the feed and expansion pipe always connected to the heating return in the position shown in Fig. 16. In the case of a combined heating and HWS installation, the feed and expansion pipe may be connected as Fig. 16 or to the primary hot-water return main close to the boiler as Figs. 18 and 20.

The method of connecting the pump to the pipework has a direct influence upon the general quietness of the installation, and for this reason a pump with flexible connections should always be choosen, or flexible connections fitted by the installer.

5

The Sizing of Radiator Connections

Radiator Connections For One-Pipe Systems

A radiator connected to a one-pipe system as shown in Fig. 17, constitutes a separate sub-circuit BCD and must be sized as such. The force producing water flow in the radiator is the sum of two quantities:

(*a*) The pressure differential $\triangle P_a$ in the ascending and descending water columns BC and CD (Fig 17) having a circulating height *h*, as in any gravity circulation (see Chapter 7 for basic information on gravity circulation).

(*b*) The pressure drop $\triangle P_b$, due to the resistance to flow in pipe BD.

From a consideration of (*b*) we deduce that the length of the pipe BD, which is governed by the length of the radiator, will determine the value of $\triangle P_b$. In a pump-operated system, the value of $\triangle P_a$ is relatively insignificant and the circulation of

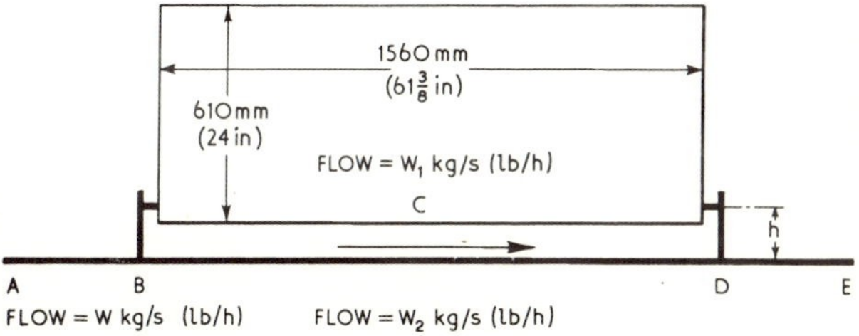

Fig. 17 *Radiator connected to one-pipe system*

water in a radiator sub-circuit BCD will mainly depend upon the pressure drop $\triangle P_b$ in pipe BD. For practical design purposes therefore, $\triangle P_a$ is ignored.

If P is the total pressure in newtons per square metre required to circulate W_1 kg/s through the radiator, and $\triangle P$ is the rate of pressure drop in newtons per square metre per metre run in pipe BD when W_2 kg/s is flowing in BD, the minimum length L of BD necessary to produce a satisfactory circulation in BCD is given by the equation

$$L = \frac{P \times 1000}{\triangle P} \text{ mm} \qquad (1)$$

In Imperial units, the above expression becomes

$$L = \frac{P \times 12}{\triangle P} \text{ inches}$$

The radiators of a one-pipe small-bore domestic system are connected to the 13 mm main branch circulating pipes with 13 mm connections (Figs. 2A, 2B and 7).

For pipe sizing, the effective length of the radiator sub-circuit BCD (Fig. 17) with 13 mm connections is 5·49 m, and, using the dining-room radiator (Fig. 2A) as an example, let us check the suitability of 13 mm connections, basing our examination on the following data.

Flow rate W for the dining-room main branch circuit, (see p. 43). $W = 0{\cdot}108$ kg/s and the flow rate W_1 for radiator sub-circuit BCD for the dining-room radiator is

$$W_1 = \frac{\text{radiator heat emission}}{\Delta t \times 4187}$$

$$= \frac{1549 \text{ (col. 10, Fig. 5)}}{11 \times 4187}$$

$$= 0{\cdot}034 \text{ kg/s}$$

then, as

$$W = W_1 + W_2$$

$$W_2 = W - W_1, \text{ and by substitution}$$

$$0{\cdot}108 \text{ kg/s} - 0{\cdot}034 \text{ kg/s} = 0{\cdot}074 \text{ kg/s}$$

which is the rate of flow in pipe BD.

From Fig. 9, the rate of pressure drop in 13 mm pipe at 0·034 kg/s is 80 N/m³, and the total pressure P required to circulate 0·034 kg/s in sub-circuit BCD will be

$$P = 80\ \text{N/m}^3 \times \text{effective length of BCD}$$

$$= 80\ \text{N/m}^3 \times 5{\cdot}49\ \text{m}$$

$$= 439\ \text{N/m}^2$$

Fig. 9 gives the rate of pressure drop in 13 mm-bore pipe at 0·074 kg/s as 330 N/m³.

Substituting the above known values in eqn. 1, the maximum length L of pipe BD, to ensure adequate pressure for circulating 0·034 kg/s through the dining-room radiator, will be

$$L = \frac{439\ \text{N/m}^2 \times 1000}{330\ \text{N/m}^3} = 1330\ \text{mm minimum}$$

Line 1, col. 9, Fig. 5, gives the actual catalogue length of the radiator as 1·56 m which means a BD pipe length of approximately 1800 mm + 100 mm = 1900 mm, thus confirming the suitability of 13 mm connections.

Where BD is found to be too short and incapable of producing the necessary pressure drop, a radiator of similar heating surface having less height and greater length will invariably provide a satisfactory solution.

Table 10 gives minimum radiator lengths in SI terms (prepared by the author) over a range of W_1 and W_2 values, to suit 13 mm connection for radiators employed with 13 mm small-bore one-pipe circuits. For Imperial units see Table 8 Appendix A. Where values of W_1 and W_2 are known, their nearest values in Table 10 will give the approximate minimum radiator length capable of producing an adequate pressure drop in BD.

Example: Consider the radiator in bedroom 3, Fig. 2B. The length of this radiator (see col. 9, Fig. 5) is 0·71 m = 710 mm, and the heat flow rate is 769 watts.

$$W_1 = \frac{769}{11^\circ\,\text{C} \times 4187} = 0{\cdot}0167\ \text{kg/s}.$$

The flow rate W for the lounge main branch circuit which

Table 10

Recommended minimum length (L) of radiator in millimetres when connected to a one-pipe 13 mm small-bore cirzuit as shown in Fig. 7 and having 13 mm connections. For intermediate flow rates (W_1) and other values of W_2, mean value of L may be computed from columns 2-9 as shown on pages 61 *et seq*.

W_1 *Radiator flow rate kg/s*	*Approximate values of W_2, kg/s*							
	0·110	0·100	0·095	0·090	0·080	0·075	0·070	0·060
(1)	(2)	(3)	(4)	(5)	(6)	(7)	(8)	(9)
up to 0·013	152	152	178	203	229	254	279	356
0·016	203	228	254	279	330	355	432	508
0·019	279	305	330	381	432	508	584	686
0·022	356	381	432	508	559	635	762	914
0·025	432	483	559	635	711	813	940	1143
0·028	533	610	686	762	864	991	1168	1397
0·031	610	686	762	864	991	1118	1321	1597
0·035	762	864	965	1067	1219	1397	1626	1957
0·038	889	991	1092	1245	1397	1597	1880	2286
0·041	1016	1143	1270	1448	1651	1880	2210	2642
0·044	1168	1321	1473	1651	1880	2134	2573	3048
0·047	1321	1473	1651	1880	2108	2413	2845	3404
0·050	1473	1626	1829	2083	2388	2692	3150	3785

serves bedroom 3 is 0·115 kg/s (see Chapter 4), and

$$W_2 = W - W_1$$
$$= 0{\cdot}115 - 0{\cdot}0167$$
$$= 0{\cdot}0983 \simeq 0{\cdot}098 \text{ kg/s}$$

Referring to Table 10 (line 1), flow values comparable with those of bedroom 3 radiator are

W_1 = 0·016 kg/s (col. 1, line 2)

W_2 = At 0·098, this lies between the Table 10 values of 0·095 and 0·100

These indicate (line 2, cols. 3 and 4) a minimum radiator length of between 228 mm and 254 mm and confirm the suitability of the proposed radiator with a length of 710 mm. The designer should always check the suitability of 13 mm connections to the radiators of one-pipe systems. Very many otherwise good installations fail to maintain room temperatures during the really cold winter periods because of sluggish water flow in the radiators.

The above remarks apply particularly to double-panel, treble-panel and column radiators, which, of course, are shorter than single-panel radiators of equal rated output. Single-panel radiators are to be preferred for one-pipe small-bore installations simply because their greater length per unit of heat emission induces a higher rate of water flow in the radiator sub-circuit BCD (Fig. 17).

Radiator Connections For Two-Pipe Systems

The radiator sub-circuit of a two-pipe system, being an extension of the main circuit, is subject to the pump circulating pressure, and the radiator connections are sized as explained in Chapters 3 and 4.

6

The Sizing, Installation and Control of the Boiler

With a combined space-heating and hot-water-supply system, boiler power is necessary to provide the installed space-heating load plus additional capacity to produce an adequate supply of hot water.

Great care should be taken, however, not to oversize the boiler. The average mean outside temperature in this country is approximately 6°C, and, as the space heating is designed for a basic temperature of – 1°C, a boiler sized to meet the space-heating load will, for most of the time, also provide all the hot water needed in the average household. During the colder periods of the heating season, when the outside conditions come nearer to the basic design temperature, the demand for space heat will increase and some extra boiler power will be needed to produce the hot water.

In Table 11 heat values are given for hot-water supply to be added to the calculated space load when sizing boilers for combined space heating and hot-water supply in houses.

Selecting the Boiler

The ability of a boiler to produce heat depends mainly upon three factors:

(1) the amount of boiler heating surface;
(2) grate area and fuel capacity;
(3) the efficiency at which the boiler operates.

If there is a correct balance between (1) and (2), the boiler being connected to a good chimney and generally well installed, then given reasonable operating attention the boiler will produce its rated heat output with ease.

Table 11

Hot-water supply heat load values to be added to space-heat requirements when sizing boilers for combined systems in small dwellings

Storage capacity of indirect cylinder (BS. 1566) litre	*Heat load allowance for boiler sizing kW*
100 – 123	2·35
145	2·93
170	3·52
214	4·69

Note: in addition to the normal hot-water requirements, the values in column 2 allow for heat emission from an insulated cylinder, or an uninsulated cylinder, in a closed cupboard. Also the heat emission from insulated or encased primary pipework not exceeding 8 m in length.

Heating boilers are rated at an average transmission of 13·9 kW/m² of boiler heating surface on the basis of a six-hour charge of solid fuel. For hot-water supply, the average rating with easy firing is 18·9 kW/m² of boiler heating surface.

The heating surface of a boiler refers to those portions of the firebox and flue surface with which the fire or the hot gases from the fire come into direct contact. Leading British manufacturers rate their boilers very conservatively and most house-heating boilers will readily develop up to 50 per cent above their catalogue rating if given proper maintenance and firing attention.

The fuel capacity of a boiler is expressed in 'cubic metres' and is the space available for fuel under actual working conditions. The heating surface and fuel capacity of house-heating boilers are given for each boiler rating in the lists of those manufacturers who are really interested in the heating performance of their products. The fuel capacity of a hand-fired boiler can be roughly measured by multiplying the length by the width of the firebox by the height measured from the top of the fire bars to the centre of the firing door.

For well-designed small house-heating boilers up to approximately 21 kW, the working fuel capacity should not be less than 0·0022 m^3 per kilowatt of boiler rating. This means that a boiler of, say, 12 kW should have a minimum working fuel capacity of $0{\cdot}0022 \times 12 = 0{\cdot}026\ m^3$.

Working fuel capacity is often confused with the total fuel capacity of the firebox which, of course, embraces both fuel and combustion space. Where a small boiler is listed as having a fuel capacity which exceeds the factor of 0·0022 m^3 per kilowatt the designer should check the manufacturer's interpretation of 'Fuel capacity.'

The above remarks apply to hand-fired standard-type heating and hot-water supply boilers, which may also be used for automatic firing with solid fuel, oil or gas. Gravity-feed boilers and also specially designed oil or gas burning boilers are not necessarily rated on heating surface, but on the maker's 'test bench' figures.

Boiler Margin

It has long been customary when selecting a boiler for a given load to choose one having a margin of 20 to 25 per cent to allow for so-called 'easy firing.' The conservative ratings given to their boilers by the best manufacturers render a margin for this purpose quite unnecessary.

A boiler margin is only justified in the case of intermittently heated buildings, or to provide overload capacity for those short periods during the winter when the outside temperature falls below the basic design temperature of -1°C. Dwelling houses unlike churches, meeting halls and similar intermittently heated buildings, are, as a general rule, constantly heated and a margin on this score is uncalled for. A margin to cover the 'below basic design periods' is also unnecessary, because during these short periods the open fires, electric or gas fires which are usually available in the living rooms of dwellings can be used to provide the small amount of heat for topping up.

Corrosion on the fire side of boilers in which the temperatures on the water side fall below the dew point of the acid combustion products, has been shown to be more pronounced in the case of oversized boilers.

Referring to oil-fired boilers, the LCC Scientific Adviser has stated: 'The results of corrosion are intensified in hot-water heating boilers because of the lower temperatures—up to 180°F (82°C) and exceptionally up to 200°F (93°C) in the water sides of the boilers. At present the only practical step to mitigate the trouble is to match the rate of firing with the heating requirements, thus reducing the cutting out of burners to the minimum.'

If the heat load has been calculated as described in Chapters 1 and 3, and provided the boiler selected to balance the heat load has the correct amount of heating surface with a firepot having a fuel capacity to match the heating surface, there is no sound reason for introducing a boiler margin.

Sizing the Boiler

The boiler for the typical installation described earlier can now be sized.

From Fig. 5, Chapter 3, the total connected heat load (exclusive of the kitchen) is 10·274 kW. The emission from the mains in the kitchen was calculated at 724 W which added to the emission from the boiler outside surfaces of 293 W makes a total of 1017 W.

The total space heating load is therefore

Heat in rooms	=	10·274 kW
Heat from mains and boiler	=	1·017 kW
Total	=	11·291 kW

Boiler heating surface for space heating

$$\frac{11{\cdot}29}{13{\cdot}9} = 0{\cdot}812 \text{ m}^2$$

The indirect hot-water cylinder for this installation will have a nominal capacity of 170 litre and Table 11 gives a boiler power of 3·52 kW.

Boiler heating surface for hot-water supply =

$$\frac{3{\cdot}52}{18{\cdot}9} = 0{\cdot}186 \text{ m}^2$$

Calculated boiler heating surface =

$$0{\cdot}812 + 0{\cdot}186 = 0{\cdot}998 \simeq 1 \text{ m}^2$$

The boiler chosen should have a guaranteed minimum heating surface of 1 m^2 and a fuel capacity under working conditions of not less than

$$0{\cdot}0022 \times 14{\cdot}81 = 0{\cdot}033 \text{ m}^3$$

Intermittently Heated Houses

As stated earlier, because most houses are occupied continuously by one or more people, a twenty-four hour heating service is necessary.

In those cases where members of the family are away at work all day, the heating system will usually be adjusted to operate at a reduced output during the unoccupied period. When this method of operation is practised it becomes necessary to add an overload allowance to the calculated heat load, (Fig. 5, col. 11 and Fig. 6, col. 12).

BS. CP 3006 : Part 1 : 1969 suggests overload margins of 10–20 per cent of the calculated building heat loss, for gas, oil and solid-fuel gravity-feed and hopper boilers, and 20–30 per cent in the case of hand-fired solid-fuel sectional or pot-type boilers.

The boiler size should not be based on the estimated building heat loss (e.g. col. 3, Fig. 6) but rather on the installed heat load (e.g. col. 12, Fig. 6) which may exceed the building heat losses by from 5 to 10 per cent.

In deciding the overload margin the designer will also consider the catalogue rating of the boiler which also rarely matches exactly the calculated total load, thus necessitating the selection of an oversized boiler.

Consider a typical case of boiler sizing for the combined system represented by Figs. 2A and 2B, the results given below are correct to two decimal places.

Heat emitted to rooms from system as installed	$= H_1 =$	10·27
Heat loss from pipes in unheated spaces	$= H_2 =$	1·02
Overload margin. 10 per cent of $(H_1 + H_2)$	$= H_3 =$	1·13
Heat for domestic hot water	$= H_4 =$	3·52
total heat load	$Q =$	15·94 kW

The boiler is selected from the following typical range of gas-fired units:

Btu/h × 2·931 × 10^{-4}	= kW
33000	= 9·67
35000	= 10·26
40000	= 11·72
45000	= 13·19
50000	= 14·66
55000	= 16·12
60000	= 17·59
65000	= 19·05
70000	= 20·52

In this case, the designer should specify a unit of 16·12 kW rated output.

After deducting the HWS heat loading of 3·52 kW, the capacity available to balance $(H_1 + H_2 + H_3)$ is

$$16{\cdot}12\ \text{kW} - 3{\cdot}52\ \text{kW} = 12{\cdot}60\ \text{kW},$$

which gives an overload based on the installed heat load, of

$$\begin{aligned} &12{\cdot}60\ \text{kW} - (H_1 + H_2) \\ &= 12{\cdot}60\ \text{kW} - 11{\cdot}29\ \text{kW} = 1{\cdot}31\ \text{kW}, \end{aligned}$$

which is equivalent to 11·61 per cent of the installed space heat load, or 13·32 per cent of the calculated building heat loss.

For hand-fired solid-fuel systems in small houses of 279 mm (11 in.) brick construction, the requirements of CP 3006 will be met by an overload margin of 15–20 per cent of the installed heat load, and 10–15 per cent for gas, oil and automatic solid-fuel systems, and for dwellings of lower thermal capacity (with insulated walls) these margins may be reduced by 5 per cent. The value of the allowance within the above limits, will depend upon the length of the unoccupied period, and the degree of output depression during that period.

A system set to give an output of 15 per cent full load from 8 a.m. to 6 p.m. will obviously need greater powers of acceleration to bring the rooms to a comfortable state in the evening, than a similar installation set to operate from 9 a.m. to 4 p.m. at 25 per cent output.

Consider a hand-fired solid-fuel installation in a house which is unoccupied on 5 or 6 days each week, between the hours of

8 a.m. and 6 p.m. The boiler will be banked before 8 a.m., and the dampers set to give a much reduced output (usually estimated at 15 per cent of full rated output) during the day. At 6 p.m. the dampers are fully opened, the furnace is cleaned, and recharged with fuel to raise the boiler flow to its maximum working temperature as quickly as possible.

In order to accelerate the heat input to the rooms without overheating the boiler, radiator and boiler capacity in excess of that required to satisfy the steady state design conditions, will be necessary.

Gas, oil and automatic solid-fuel boilers are more responsive to time and temperature control. Unlike the standard hand-fired solid-fuel boiler, these boilers may be set to give any desired output between minimum and maximum, and may also be time-switched to operate at full output for a predetermined period (termed the *preheating period*) before the owner and his family come home in the evening. For this reason the overload allowance for systems fired by automatic boilers, or by electric heating systems, is lower than for hand-fired boilers.

If the system we are considering is to be solid-fuel hand-fired, and installed in a house which is unoccupied from 8 a.m. to 6 p.m., a 20 per cent overload allowance will be appropriate. This allowance will be equal to

$$\frac{20 \times 11{\cdot}29 \text{ kW}}{100} = 2{\cdot}26 \text{ kW}$$

Both the boiler rating, and radiator surfaces must be increased to give this extra performance at the design boiler temperatures of 82°C flow and 71°C return.

The pump duty will also need amending to cope with the additional load.

Boiler Connections

Among the more serious troubles encounted with small-bore heating installations is the existence of counterflow between the gravity and forced circulations at the return connection to the boiler. This occurs when the pumped return from the space heating and the gravity hot-water supply return are connected with a tee or twin elbow into the same boiler return tapping. Owing to the pressure differential at this point there is a tendency for the lower pressure gravity circulation to be reversed. Another

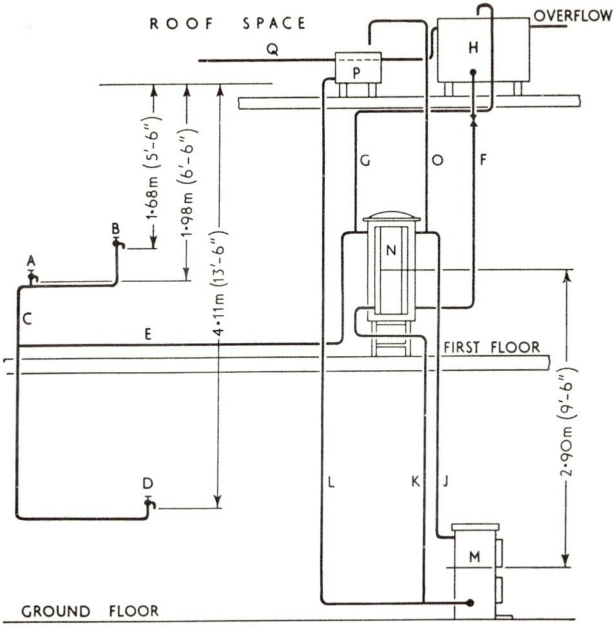

Fig. 18 *Indirect hot-water supply system*

Key

A *HWS to bath*
B *HWS to lavatory basin*
C *HWS to bath and lavatory basin*
D *HWS to sink*
E *HWS secondary main pipe*
F *Cold feed pipe*
G *Secondary open vent pipe*
H *Cold-water storage tank*
J *HWS primary flow pipe*
K *HWS primary return pipe*
L *Primary feed and expansion pipe*
M *Heating and HWS boiler*
N *170 litre indirect cylinder*
O *Primary open vent pipe*
P *Feed and expansion tank*
Q *Cold-water service pipe*

contributing factor is an oversized pump which increases the pressure differential unnecessarily.

The only practical solution is to ensure that a pump correctly matched and regulated to the index circuit is installed, together with a boiler having two return tappings, one on either side, the primary hot-water return being connected to one tapping and the space-heating return to the tapping on the opposite side. Small house-heating boilers with two flow and two return tappings are easily obtainable and only those boilers equipped in this way should be specified for combined small-bore heating and hot-water supply systems.

With gravity-operated combined systems this trouble does not occur because the pressure differential between the gravity circulated heating and primary hot-water circuits is so much less than in small-bore systems.

A bulletin published by the Heating and Ventilating Research Association (Technical Note No. 6, 1959. Billington, N.S.) suggests a method of connecting a combined gravity primary return,

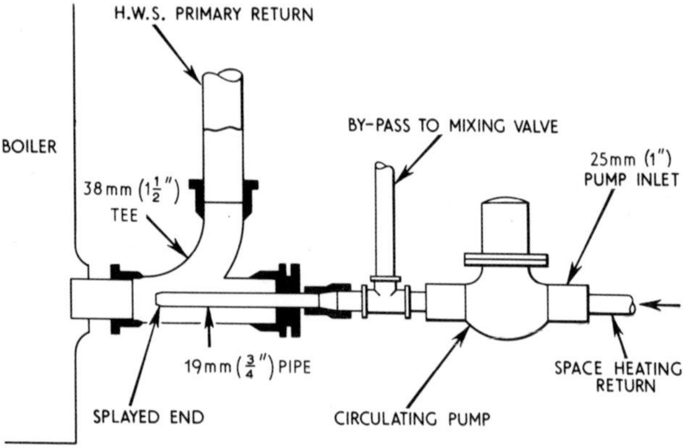

Fig. 19 Method of connecting hot-water primary return and small-bore heating return to single-boiler return tapping

and a pumped small-bore heating return into a single boiler return tapping. Fig. 19 illustrates this method of connection, which is intended to produce an induced effect from the pump. This method, to be effective, requires an oversized primary circuit, which not only provides the necessary mixing chamber within the oversized sweep tee, but also ensures that the friction loss of the primary circuit is reduced to a minimum.

Boiler Mountings

Every boiler should be fitted with the following mountings:

one 19 mm national enclosed spring-pattern safety valve;
one thermometer;
one altitude gauge (or a combined altitude gauge and thermometer);
one 13 mm empty cock with hose connection;
one 19 mm open pipe.

The open pipe may be connected to the primary flow to the cylinder as shown in Figs. 16 and 18. Valves must never be fitted in the primary flow and return pipes, and the safety valve should always be fitted into a boiler tapping and not in any other position or pipe.

The safety valve should be loaded to open when the pressure in the system in newtons per square metre (N/m^2) is equal to

$$P = (\text{static head of system} + 3\ \text{m}) \times 9807$$

The static head is read from the altitude gauge, and an example will show the application of this formula.

Suppose the static head is shown to be 5 m then the correct safety-valve setting will be

$$P = (5\ \text{m} + 3\ \text{m}) \times 9807$$
$$= 8 \times 9807$$
$$= 78456\ \text{N/m}^2$$

which is approximately equal to

0·785 bar

Automatic Control

Automatic control is necessary in the interests of fuel economy, and to ensure reasonably constant room temperatures and an adequate hot-water supply despite variable weather conditions. The boiler temperature of a combined system is dictated by the heat demands of the hot-water service. Hot water for domestic purposes should be delivered to the taps at 60°C which, allowing for a temperature drop at the tap farthest from the cylinder, means that the hot water in the cylinder should be maintained at or near 66°C. To produce water at this temperature in an indirect system, the boiler temperature must never fall much below 82°C.

For space heating, the water temperature is required to vary according to changing demands of the weather, and water temperatures considerably below 82°C are often adequate to meet the load. There are two ways in which these different heat demands can be met from the one boiler. These are:

(1) to operate the boiler at a constant temperature of 82°C to ensure an adequate hot-water supply at all times, and to control the heating pump by a room thermostat.

(2) alternatively, to operate the boiler at a constant temperature of 82°C and install a return bypass and mixing valve arrangement to enable the flow water for the space-heating system to be varied in temperature, by mixing boiler water at 82°C with return water from the space heating in proportions to suit the demands for space heat at any time.

The simplicity of method (1) has much to commend it, but in practice is not very successful. Where systems are controlled in this way it is found that with the pump stopped there is a tendency for gravity circulation to occur on the first floor, allowing the room temperature to rise unnecessarily causing discomfort and a waste of fuel. Control of a house-warming system by one room thermostat is rarely sucessful, owing to the difficulty of choosing the room for the thermostat. Suppose the lounge is the room chosen, if the open fire is lit, an electric fire switched on, a fortuitous heat gain from a party held in the room, or solar heat raises the room temperature, the thermostat will switch off the whole heating system. Even when an intermittently occupied room such as the dining room or a bedroom is chosen, the control remains effective only provided the radiator in that room is continually turned on, which again leads to discomfort, fuel wastage and increased running costs.

The ideal application of method (2) is to operate a mixing valve through a temperature-sensitive element fixed on an outside wall, the element being connected to the mixing valve by a capillary tube or by an electrical circuit. The outside element controls the hot and return water ports of the mixing valve to give the correct heating flow temperature. In addition, two immersion thermostats are fitted to the boiler, one of which operates as a running control to close the dampers or switch off the oil burner when the boiler water temperature rises above 82°C, and to reverse the process should the temperature fall below 82°C. The second thermostat is a boiler limit control and is set at 6°C above the normal working temperature to operate should the running thermostat fail to function, and so prevent over-heating and damage to the installation.

A cheaper but not quite so effective application of method (2) has been generally adopted for combined hot-water and small-bore heating. This dispenses with the outside control element and employs an automatic three-way mixing valve with an internal thermostatic element which regulates the opening of the hot and cold water ports of the valve to provide the correct mixed water temperature at the valve outlet. The mixed water temperature is indicated on a dial on the valve face and, to enable temperature adjustments over the normal operating range of the heating system to be made, a manual overriding control is provided. Once the manual control is set, the bi-metal thermostat within the control valve will adjust the flow temperature to meet the varying demands of outside climatic conditions.

Because of its simplicity and relative cheapness, this system of control has been widely adopted and is most suitable for

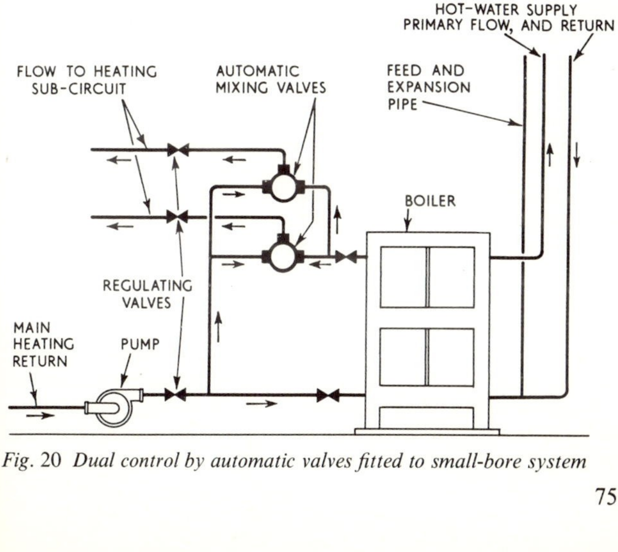

Fig. 20 Dual control by automatic valves fitted to small-bore system

house-heating installations. Perhaps the greatest defect in this application of the mixing valve method is that unlike the outside element application it is not sensitive to variations in wind velocity and direction. This can become important on an exposed site when, at certain times of the year, the effect of strong, cold winds may call for an increase of up to 6°C in the mixed water temperature. A short-coming of this nature is much more critical in large multi-storey buildings than in small dwelling houses however and, although the three-way automatic mixing valve as generally applied is not a perfect control, it does enable the diverse load requirements of domestic hot water and house warming to be provided by one boiler with reasonable effectiveness and economy.

The fixing position of the three-way automatic mixing valve is shown at H in Fig. 8 (Chapter 4) and Fig. 20 illustrates the application of the valve for the control of individual heating sub-circuits.

Room Temperature Control

Although the central control described will regulate the flow water temperature for the whole system, it will not give precise temperature control of individual rooms. With small-bore systems control of room temperature is made possible by the use of thermostatic radiator valves. A valve of this type can be fitted to the inlet of each radiator with the usual lockshield valve on the outlet. Rooms receiving heat gains from other rooms or from open fires or other fortuitous sources within the rooms, and also those rooms having a warm orientation, which even in the winter may receive appreciable solar heat gain, should be considered for thermostatic radiator control.

Because of its very low water content, the small-bore system responds quickly to thermostatic radiator control with a markedly good effect upon both room comfort and fuel consumption.

Manual regulation of water flow through the radiator while not so effective as thermostatic control will give extremely good results, provided each radiator inlet is fitted with a specially-designed glandless regulating-valve. These valves are available in both angle and straight patterns, with a dial setting which indicates the degree of valve opening and enables the radiator output to be controlled within remarkably close limits. Although not so selective as the thermostatic radiator valve the manually-operated regulating valve is cheaper to install, and if intelligently used by the house-holder will contribute much towards the comfort and economy of the installation. It is in the interests of both the installer and the householder that the latter should understand the correct operation of his heating and hot-water system.

Time-Switch Control

The time switch is being increasingly used for effecting operating economies in house warming, and is useful for small-bore systems where gas, electric, oil or automatic solid-fuel boilers are used. The time switch, which is used in conjunction with the thermostatic controls previously discussed, is wired to operate the boiler and the circulating pump simultaneously, or separately.

During spring and autumn when the heating demand is reduced, the time switch may be set to operate the pump to provide limited space heating, say, during the morning and evening, leaving the boiler on continuous duty for normal hot water supply.

Where the occupants are out at work all day, the time control unit can be set to switch on both the boiler and pump to ensure a good supply of hot water and a warm house when the people come home in the evening. During the summer when the space heating is shut down, the time control is set to operate the boiler only to provide hot water for early morning and evening use.

For the dwelling which is occupied twenty-four hours each day, the time switch will effectively control the running periods of the installation to suit the particular requirements of the household.

Packaged boiler and pump control units are available which give selective switchings, enabling the space heating and the hot-water supply to be programmed to meet the needs of the household, and the seasonal demand.

Complete automatic time and temperature control undoubtedly produces the most economical results from the point of view of fuel consumption. It must be remembered, however, that the more working parts there are, the more maintenance is necessary to keep the installation working at peak efficiency.

In complicated control mechanisms the failure of one small component will put the whole heating system out of action until remedied by the manufacturer, who alone properly understands the mechanism and its maintenance.

For systems in small houses which are continuously occupied, and where precise control involving intermittent operation is not an important factor it is, therefore, wise to adopt a simple arrangement, with a measure of manual control, rather than numerous automatic time and temperature controls which are liable to break down at any time. Where fully automatic time and temperature control is specifically required, the advantages and disadvantages of such a system should be carefully assessed.

Zonal Control

Where the installation consists of two or more main branch circuits, as in the examples given in this book, zonal control of room temperature, to supplement the main boiler water temperature control, may be considered, and will in suitable cases, and especially where houses of severe or normal exposure are concerned, save sufficient fuel to justify fully the additional capital cost involved.

In the example, Figs. 2A and 2B, the dining-room circuit serves rooms having a westerly or south westerly orientation, and the lounge circuit those rooms having an easterly orientation.

Zonal control may be achieved by fitting a 13 mm motorised valve in the flow branch of each circuit, to be actuated by room thermostats. The thermostat for the west zone may be fixed in the dining room, and that for the east zone in the lounge or the hall.

With zonal control, the room thermostat in the coldest zone should operate both the motorised valve and circulating pump.

Domestic Hot-Water Temperature Control

A selection of thermostatically operated valves designed to control the water in an indirect cylinder at a predetermined temperature is available. Basically, this type of control consists of a thermostat immersed in the secondary water in the cylinder, which actuates a valve in the primary circuit.

The use of this type of control should be restricted to those small-bore combined systems which are automatically fired by gas, oil or electricity, and which, as explained in Chapter 7, are much less prone to overheating than hand-fired solid-fuel boilers.

7

Sizing the Hot-Water-Supply Piping System and the Cold-Water Storage

In Chapter 6, the indirect cylinder size was given as 170 litre nominal capacity, and in sizing the boiler a heat output of 3·52 kW equal to 0·186 m^2 of boiler heating surface was allowed. These sizes can be regarded as the minimum standard for combined systems in the small type of houses which constitute the great majority of dwellings in this country.

Sizing the Primary Flow and Return Pipes

The purpose of the primary circuit is to transmit the heated water from the boiler to the inner cylinder via pipe J (Fig. 18). In the inner cylinder it gives up heat to the water in the outer cylinder before returning in pipe K to the boiler. Circulation must proceed at a rate which will enable the water in the outer cylinder to receive all the heat which the 0·186 m^2 of boiler is capable of imparting to it, and consequently the primary flow and return pipes should be sized to circulate 3·52 kW

The value of W, the water flow rate in the primary circuit MJNK Fig. 18, is given by

$$W = \frac{\text{heat flow in watts}}{\Delta t \times 4187}$$

Where W is the flow in kilogrammes per second, and Δt is the temperature difference in degrees C between the water in pipes J and K, Fig. 18.

In an HWS primary circuit, Δt is not constant and will be at a maximum value when the draw-off from the system is high, falling to its minimum value when little or no draw-off is taking place. A Δt value of 16° C has been found to represent average conditions.

Water will leave the boiler via pipe J at 82°C and, after giving up its heat to the water in the cylinder, will return at 66°C via pipe K to the boiler.

Unlike the space-heating circuit, the primary circulation is by gravity. Circulation by gravity depends upon the density (or weight) of the water in the vertical pipe J (Fig. 18) being less than the density of the water in the vertical pipe K. As the water in M and J becomes heated expansion occurs, and its density decreases, and, obeying the law of gravity the heavier water in K descends to M displacing the warmer and lighter water, forcing it out of M, up pipe J, and into the cylinder N where it again cools, the process being repeated over and over again.

In Imperial units, the force producing circulation, or the circulating pressure in feet, per foot in height of the circuit, is found by dividing the difference in density of the water in the return pipe K and flow pipe J, by the mean density of the water in J and K.

		lb/ft³
Density of water at 150°F	=	61·201
Density of water at 180°F	=	60·560
Difference in density	=	0·641

Mean density of water in J and K

$$= \frac{60{\cdot}560 + 61{\cdot}201}{2} = 60{\cdot}880 \text{ lb/ft}^3$$

The circulating pressure in inches water gauge, per foot in height of the primary circuit will be:

$$\frac{0{\cdot}641 \times 12 \text{ in.}}{60{\cdot}880} = 0{\cdot}126 \text{ in. w.g.}$$

In SI units, the circulating pressure is expressed in newtons per square metre, and repeating the above calculations in metric terms we have

		kg/m^3
Density of water at 65·6°C	=	980·17
Density of water at 82·2°C	=	970·40
Difference in density	=	9·77

Mean density of water in J and K

$$= \frac{980{\cdot}17 + 970{\cdot}40}{2} = 975{\cdot}285 \text{ kg/m}^3$$

From which we may calculate the circulating pressure in millimetres H_2O per metre of circulating height as

$$\frac{9{\cdot}77 \times 1000}{975{\cdot}285}$$

In the SI system the unit of pressure is the newton per square metre (N/m^2), and 1 mm $H_2O = 9{\cdot}807$ N/m^2. Therefore the circulating pressure in newtons per square metre per metre (N/m^3) of circulating height of the gravity circulation MJNK will be

$$\frac{9{\cdot}77 \times 1000 \times 9{\cdot}807}{975{\cdot}285}$$

$$= \frac{9{\cdot}77 \times 9807}{975{\cdot}285} = 98{\cdot}2 \text{ N/m}^3$$

This may be called the difference in density between the flow and return columns of water, multiplied by a constant of 9807, and divided by the mean density. The total circulating pressure available is found by multiplying the pressure per metre in height (in this case 98·2 N/m^3) by the circulating height in metres. Circulating height is measured from the centre of the boiler flow and return tappings, to the centre of the cylinder primary flow and return tappings. In Fig. 18 this distance is 2·90 m, and the total available circulating pressure will be

$$CP = 98{\cdot}2 \text{ N/m}^3 \times 2{\cdot}90 \text{ m}$$

$$= 285 \text{ N/m}^2$$

Pressure in a circuit is necessary to overcome:

(1) frictional resistance to movement between the water, and the internal surfaces of the pipes;

(2) the resistance to flow created by elbows, bends, valves, boiler, cylinder and other fittings in the circuit, which are known as 'local resistances.'

The chart (Fig. 9) gives the pressure drop due to friction in small copper pipes for different rates of flow, and Table 9 shows the values of local resistances in equivalent lengths of straight pipe. For steel pipes of equal bore, the pressure drop is increased by 12 per cent.

Example Size the primary flow and return pipes as depicted in Fig. 18 which represents a typical arrangement for a small dwelling house.

The heat flow rate for hot-water production is 3·52 kW = 3520 watts, and the mass flow rate in the primary circuit will be

$$W = \frac{3520}{(82^\circ\text{C} - 66^\circ\text{C})\,4187}$$

$$= \frac{3520}{16 \times 4187}$$

$$= 0{\cdot}05254 \simeq 0{\cdot}053 \text{ kg/s}$$

In small houses the size of the primary circuit will be within the range of 19–32 mm bore, depending upon the effective length of the circuit, and the circulating height, or, in other words, depending upon the relative positions of the boiler and cylinder; in the present case assume a provisional pipe size of 25 mm (1 in).

The measured length of the circuit of Fig. 18 is 7·62 m to which must be added the equivalent length for local resistances, which are

From Table 9

Boiler	1 × 3·66 m	= 3·66 m
Cylinder	1 × 1·62 m	= 1·62 m
1–25 mm through tee	1 × 0·12 m	= 0·12 m
1–25 mm tee (flow into)	1 × 1·37 m	= 1·37 m
5–25 mm bends	5 × 0·58 m	= 2·90 m
Equivalent pipe length for local resistances		= 9·67 m

The effective length of the primary circuit for pipe-sizing purposes will be

Measured length of circuit	=	7·62 m
Add for local resistances	=	9·67 m
Total effective length	=	17·29 m

The total available pressure is 285 N/m² and the maximum allowable pressure drop per metre run of the circuit will be

$$\triangle P/\mathrm{m} = \frac{285\ \mathrm{N/m^2}}{17{\cdot}29\ \mathrm{m}} = 16{\cdot}48\ \mathrm{N/m^2}$$

$$\simeq 16{\cdot}5\ \mathrm{N/m^2}$$

From Fig. 9 a 25 mm pipe will circulate 0·053 kg/s with a pressure drop of 7 N/m² per metre run, which multiplied by the effective length of the circuit gives a total head loss of

$$7\ \mathrm{N/m^2} \times 17{\cdot}29\ \mathrm{m} = 121\ \mathrm{N/m^2}$$

This is considerably below the working pressure available of 285 N/m², and a 19 mm pipe may be satisfactory. The possibility of installing a 19 mm primary flow and return will be examined. Measured length of primary circuit Fig. 18 = 7·62 m. Add local resistances taken from Fig. 18 and Table 9:

Boiler	1 × 2·74 m	= 2·74 m
Cylinder	1 × 1·13 m	= 1·13 m
1 – 19 mm through tee	1 × 0·08 m	= 0·08 m
1 – 19 mm tee (into branch)	1 × 0·98 m	= 0·98 m
5 – 19 mm bends	5 × 0·40 m	= 2·00 m
Equivalent length of pipe for local resistances		= 6·93 m

The effective length of the circuit for pipe-sizing purposes will be

Measured length of circuit	=	7·62 m
Add for local resistances	=	6·93 m
Total effective length of circuit	=	14·55 m

From Fig. 9 a 19 mm pipe will circulate 0·053 kg/s with a pressure drop of 26 N/m² per metre run, which, when multiplied by the effective length of circuit, gives a total head loss of

$$26\ \text{N/m}^3 \times 14{\cdot}55\ \text{m} = 378{\cdot}3\ \text{N/m}^2$$

$$\simeq 378\ \text{N/m}^2$$

This is much greater than the available pressure of 285 N/m^2, and indicates the inadequacy of a 19 mm flow and return. 25 mm flow and return pipes should be installed.

The Danger of Overheating in Small-Bore Systems

Under certain operating conditions with a hand-fired solid-fuel combined small-bore heating and hot-water supply system, an oversized primary circuit is advantageous. There are occasions when the space-heating side of the installation may, owing to a rise in the outside temperature, cease to call for heat from the boiler. Should this occur when the boiler fire-pot is full of incandescent fuel, the residual heat of the furnace will force up the boiler water temperature even though the dampers may be closed. If the cylinder is not already full of hot water it will accept part or the whole of the heat, and so prevent overheating of the system and damage to equipment.

In such circumstances an oversized primary circuit is advantageous and for this reason alone the primary flow and return should never be undersized.

In the case of direct hot-water systems where scaling up is likely to occur, it is also better for the primary circuit to be oversized, because even an extremely thin layer of scale upon the internal surfaces of the pipes will greatly increase the frictional resistance.

The danger of overheating with solid-fuel boilers, and also with gas- and oil-fired boilers operating on the high/low/off principle of automatic combustion control, is diminished by the gravity circulation to the hot-water storage cylinder. As an added precaution, connect the bathroom radiator, and where convenient, the landing radiator to the cylinder gravity circulation.

When this is done, the primary flow and return from the boiler should be increased to 32 mm bore, and tees inserted in the primary flow and return to enable 19 mm two-pipe radiator connections to be taken to the bathroom or the landing radiator, or both.

Correct, and incorrect methods of connection are shown in

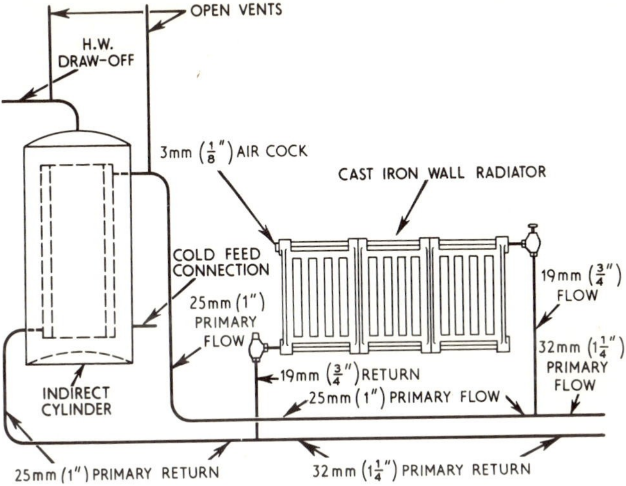

Fig. 21 Radiator gravity circuit to relieve overheating; correct method

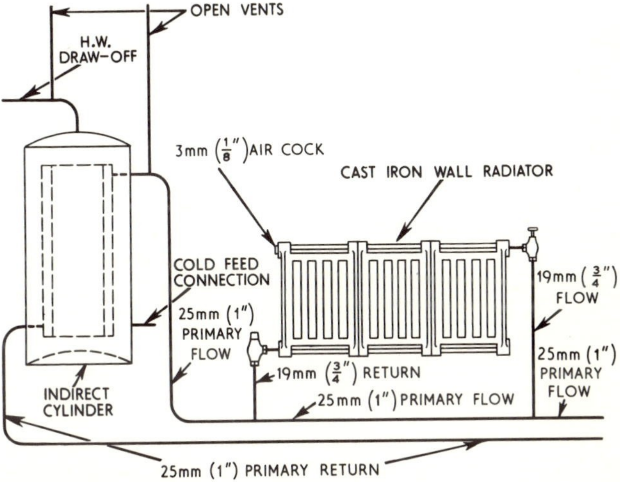

Fig. 22 Radiator gravity circuit to relieve overheating; incorrect pipe sizing and wrong method of connection

Figs. 21 and 22. Fig. 22 shows the incorrect way which is often adopted for convenience, and which reduces the hot-water output owing to the cooling effect of the radiator.

The designer must exercise great care to ensure an adequate gravity circulation to prevent overheating in the event of a pump stoppage for any reason.

Gas, oil and electric boilers operated by automatic on/off controls quickly respond to the limit thermostat in the event of a pump stoppage, and the danger of heat build-up is therefore very much reduced.

Sizing the Secondary Pipes

The secondary piping system embraces all the pipework from the cold-water storage tank to the taps, its purpose being to convey the hot water in sufficient quantity to the taps, at the required temperature.

Consider Fig. 18. When taps A, B and D are closed, the water in tank H, cold feed pipe F, cylinder N, and secondary pipes B, C, D and E is motionless. If tap A, B or D is opened, hot water will flow along E from N, and cold water will flow into N via F from the cold-water storage tank H. The ball valve in H will open, and cold water will flow into H from the cold-water service pipe Q. Provided sufficient water can flow from Q to maintain a constant level in H, the rate of flow from the open hot-water tap will also be constant. The velocity of the water leaving the tap will be that due to the static head as measured from the water level in H to the tap outlet, less the head absorbed in pipe friction and by local resistances.

If the pipes are so long and the fittings so numerous, or the pipes so small as to absorb more than the total head available, the outlet velocity and the amount of water leaving the tap will be small, and the installation will have failed in its purpose. The amount of water which will flow from a hot-water system depends therefore upon the height of the cold-water storage tank above the taps, and upon the frictional resistance of the pipes and fittings through which the water flows on its way to the taps.

It is not possible to guarantee that the cold-water storage tank will remain full at all times, especially during the day-time peak demand hours when the town water pressure may be at its lowest. For this reason, for practical design purposes, the static head should be measured from the bottom of the cold-water tank to tap outlet. The pipes must be sized so that their frictional resistance and local resistances balance or nearly balance the static head when the system is delivering the correct quantity of water to the taps.

Fig. 23 gives the head loss due to frictional resistance per metre length for copper pipes of sizes usually required for domestic water services in small dwellings. To enable the loss of head caused by local resistances to be calculated, Table 9 gives

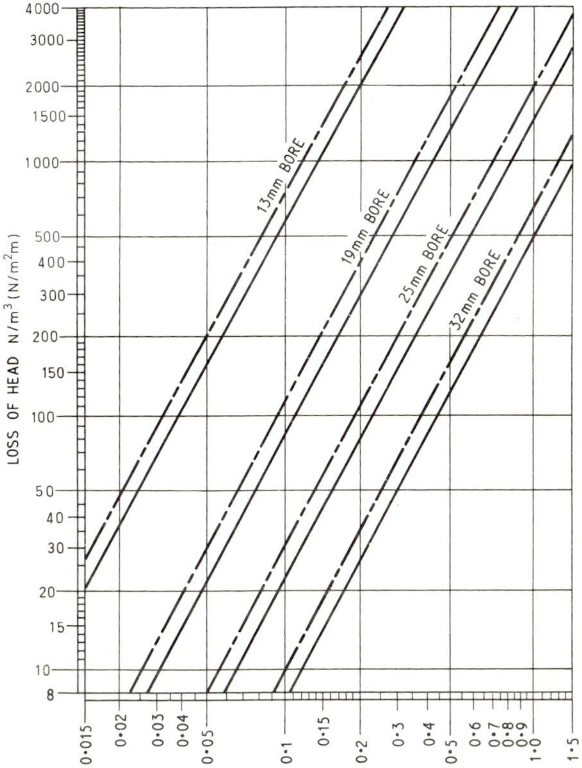

Fig. 23 Flow of water in litres per second at 65·6°C and 10°C in copper pipes; BS 659 : 1955

——————— *at 65·6°C*

— · — · — *at 10°C*

the equivalent length of straight pipe for each resistance.

The Rate of Flow

The rate at which water should flow in a pipe is that which will provide the required discharge at the taps. Pipework for domestic hot water should be designed to allow the following rates of flow in litres per second:

Bath	0·38
Lavatory basin	0·11
Sink	0·30
Domestic wash boiler	0·30
Domestic washing machine	0·30

Probable Highest Simultaneous Demand (PHSD)

In a small dwelling, the hot-water installation serves a bath, a lavatory basin (abbreviated LB, and sometimes called a wash basin), and a kitchen sink, the two first-named being fixed side-by-side in the bathroom.

Under normal living conditions, the simultaneous use of the bath and LB is unlikely, more practical combinations being the bath and sink to produce a simultaneous demand of 0·38 litre/s + 0·30 litre/s = 0·68 litre/s, or the sink and LB, when the simultaneous demand is 0·30 litre/s + 0·11 litre/s = 0·41 litre/s.

For a small dwelling the PHSD will therefore be 0·68 litre/s and the main secondary flow pipe (pipe E, Fig. 18) and the cold-water feed pipe (pipe F) are sized for this flow rate.

PHSD values for the secondary pipes of Fig. 18 are

Pipe A to bath	= 0·38 litre/s
Pipe B to LB	= 0·11 litre/s
Pipe C serving pipes A and B	= 0·38 litre/s
Pipe D to sink	= 0·30 litre/s
Pipes E and F serving pipes ABC and D	= 0·68 litre/s

Secondary Pipe-Sizing Procedure

Fig. 24 illustrates a rational procedure for secondary hot-water pipe sizing, in which the secondary system (Fig. 18) is divided into three sections which are identified by the pipe references, as

Section BCEF serving the LB

Static head = 1·68 m H_2O
= 1·68 m H_2O × 98·07 = 165 mb (16500 N/m^2)

Section ACEF serving the bath
Static head = 1·98 m H_2O
= 1·98 m H_2O × 98·07 = 194 mb (19400 N/m^2)

Section DEF serving the sink
Static head = 4·11 m H_2O
= 4·11 m H_2O × 98·07 = 403 mb (40300 N/m^2)

Secondary pipes for small domestic systems come within the 13 mm (½ in) 32 mm (1¼ in) range, the provisional sizes (column 3 of Fig. 24) are chosen from this range, and the calculations and data in the columns which follow column 3 will determine the suitability of the provisional sizes, this being confirmed or otherwise in column 11 of Fig. 24.

Pipes smaller than 13 mm bore (½ in) are not recommended for domestic hot- and cold-water installations. Sizing commences with the 'least favoured' or 'index' draw-off, which, in the scheme of Fig. 18 is pipe B, the sizing process being recorded in lines 1 to 4 of the scheme sheet Fig. 24.

Beginning with line 1, the procedure is:

Column 1 Pipe reference: B
Column 2 Water flow rate: 0·11 litre/s
Column 3 Provisional pipe size: assume 13 mm
Column 4 Length from junction with pipes C and A . . . 1·22 m
Column 5 Equivalent length for local resistances. Note the number of resistances in columns 12–18 inclusive, and using Table 9 estimate the equivalent length at 4·78 m, (column 19).
Column 6 The effective length for pipe sizing will be column 4 + column 5
= 1·22 m + 4·78 m = 6·00 m
Column 7 From Fig. 23 the head loss per metre run of 13 mm pipe for a flow rate of 0·11 litre/s, at 65·6°C = 640 N/m^3
Column 8 Total head loss in pipe B = column 6 × column 7
= 6·00 m × 640 N/m^3
= 3840 N/m^2
Column 9 Records progressively the head loss in the section
Column 10 Records the pressure head available.

Pipe reference	*litre/s*	*Provisional pipe size mm*	*Measured length m*	*Equivalent length m*	*Effective length (col. 4 + col. 5) m*	*Head loss N/m³*	*Head loss N/m²*	*Total head loss N/m²*	*Pressure head available N/m²*	*Final pipe size mm*
(1)	(2)	(3)	(4)	(5)	(6)	(7)	(8)	(9)	(10)	(11)
B	0·11	13	1·22	4·78	6·00	640	3840	3840	16500	13
C	0·38	19	0·91	0·40	1·31	900	1179	5019	(0·165 bar)	19
E	0·68	25	4·27	4·27	8·54	660	5636	10655		25
F cold feed	0·68	25	2·44	3·39	5·83	860	5114	15769		25
A	0·38	19	0·30	8·45	8·75	840	7350	7350	19400	19
C E F							11929	19279	(0·194 bar)	
D	0·30	19	2·74	8·12	10·86	580	6300	6300	40300	
E F							10750	17050	(0·403 bar)	
Revise										
D	0·30	13	2·74	4·98	7·72	3700	28564	45614		13
				Deduct line 7			6300	39314		

Pipe reference	Local Resistances							
	Valve fullway	Tap	Cylinder or tank	Elbow	Bend	Tee branch	Tee through	Equivalent length m
(1)	(12)	(13)	(14)	(15)	(16)	(17)	(18)	(19)
B		1			1		1	4·78
C					1			0·40
E			1		2	1	1	4·27
F cold feed	1		1		2			3·39
A		1				1		8·45
C E F								
D		1			2			8·12
E F								
Revise								
D		1			2			4·98

Fig. 24 *Domestic heating scheme sheet: secondary hot-water pipe sizing (refer to Fig. 18)*

Notes

Column 2 is correct to two significant figures

Columns 4, 5 and 6 are correct to two decimal places

Columns 8, 9 and 10 are correct to the nearest whole number

**In SI practice, operating pressures may for convenience be stated in bars. e.g. the pressures in column 10 become 0·165 bar, 0·194 bar and 0·403 bar respectively.*

These simple calculations are repeated in line 2 for pipe C, to give a head loss of 1179 N/m^2 (column 8) and a total head loss in pipes B and C of

$$3840 + 1179 = 5019 \text{ N/m}^2 \text{ (column 9)}$$

Repeating in lines 3 and 4 for pipes E and F the head loss for section BCEF totals 15769 N/m^2 (column 9) which is less than the pressure available of 16500 N/m^2 (column 10). For section BCEF the provisional pipe sizes are confirmed and the final pipe sizes are entered in column 11.

The sizing of section ACEF which serves the bath is given in lines 5 and 6, the total head loss based on the provisional sizes (column 3) being 19729 N/m^2 (column 9) compared with the available pressure head of 19400 N/m^2 (column 10) which confirms the required size of pipe A as 19 mm (column 11).

In sizing pipe D the sink draw-off, lines 7 and 8 show that for a provisional size of 19 mm, the head loss in section DEF is only 17050 N/m^2 compared with the pressure available, which is 40300 N/m^2 (column 10). Reducing pipe D to 13 mm (lines 9 and 10) increases the resistance of section DEF to 39314 N/m^2 giving a much closer balance with the available pressure head.

Pipes E and F are sized to ensure a good flow to the draw-offs on the first floor, and cannot therefore be reduced in size.

General Observations on Pipe Sizes for Domestic Hot-Water Installations

The pipe sizes which have been determined by the foregoing calculations should be regarded as the minimum for good hot-water practice in small houses.

The following minimum sizes are recommended for other component parts of the system:

Open vent pipes. (G and O, Fig. 18) should be 19 mm.

Feed and expansion pipe. (L, Fig. 18) should be 13 mm.

Tank overflow pipes should be 25 mm minimum and 32 mm where possible.

Feed and expansion tank for primary hot water, and heating circuits, 45 litre nominal capacity.

Sizing the Cold-Water Storage Tank

The purpose of the cold-water tank is to feed the hot-water system, to provide some storage for emergency, to allow for expansion, and to enable the installation to operate at a safe, reasonable, and fairly constant pressure.

There is no precise law or formula to guide one in the correct sizing of the tank, and perhaps the best approach to the problem is on the basis of the number of people to be catered for.

Houses of the type we are considering usually have two double and one single bedrooms, which under normal living conditions means accommodation for five people.

The following minimum amounts of hot water at 60°C are used for domestic purposes:

Bath (adult)	68·0 litres
Lavatory basin used for strip wash	7·0 litres
Lavatory basin used for washing hands	4·5 litres
Lavatory basin used for shaving	2·3 litres
Sink for washing dishes	36·0 litres
Domestic washing machine or copper	45·0 litres

In the average household the maximum amount of hot water is used at the weekend. Consider the probable hot water usage during Sunday in a household of five adults. Upon rising each person will either strip wash or take a bath. In any case it is reasonable to estimate that sometime during the day all the occupants will strip wash and in addition four of the five will have a bath. There will also be occasional washing of hands, shaving, food preparation and washing dishes. Clothes washing and house cleaning will in the normal household be confined to weekdays.

Based upon this reasoning the probable hot-water usage on Sunday, the day of heaviest demand, will be:

4 baths. 68 litres × 4	= 272 litres
5 people strip wash. 7 litres × 5	= 35 litres
2 people shave. 2·3 litres × 2	= 4·6 litres
5 wash hands twice. 4·5 × 5 × 2	= 45 litres
Food preparation. 36 litres	= 36 litres
Washing dishes after 4 meals. 36 litres × 4	= 144 litres
Estimated total hot-water consumption	= 536·6 litres

This represents a daily consumption per person of 107·3 litres and 110 litres per person per day is a safe basis for calculation, and will allow a margin to cover additional consumption which occurs when guests are entertained.

Where a good water supply at reasonable pressure exists a storage of approximately a half day's consumption is sound practice. If the cold-water service is uncertain owing to low or widely fluctuating pressure, or liable to frequent break-down, a full day's consumption should be stored. In mining areas subject to subsidence and the resultant damage to underground mains, the water authorities insist upon a full day's storage.

For a three-bedroom house the cold-water storage for hot-water purposes will be:

Five persons × 110 litres = 550 litres for a full day's supply, and say 275 litres to provide sufficient storage for half a day. Nominal tank sizes to ensure these amounts should be installed.

8

Estimating the Annual Fuel Cost

The domestic heating engineer is often called upon to advise a prospective customer on the choice of fuel. When asked which is the best fuel to use for domestic heating, we can quickly name the cheapest fuel as delivered, or the cleanest fuel to use. When, however, we carefully examine the annual operating costs of installations using the different fuels: coke, anthracite, oil, gas and electricity, and also consider their individual operating characteristics, we are forced to the conclusion that no one fuel has an overwhelming advantage and that all are competitive.

With the growing popularity of domestic central heating, there is also an increasing number of discerning house owners who are inclined to weigh carefully the advantages of cleanliness and full automatic operation of one fuel against the purely monetary advantage of others. The merits and demerits of the various fuels are, generally speaking, well appreciated by the layman, and usually the main purpose of a customer's enquiry is to be advised on the annual cost of his proposed installation.

Annual cost is made up of day-to-day running costs such as fuel, maintenance and repairs, flue cleaning, and electricity for operating motors and controls, and also standing charges which cover the purchase of the installation, insurance and incidental items, e.g. cost of redecoration of room where a boiler is fixed. The last item is included because rooms where boilers are fixed need decorating much more often than rooms without boilers.

Estimating Fuel Consumption

The amount of fuel consumed depends upon several factors: efficiency, quality of design, quality of maintenance, quality of operational attention, the weather, the length of the heating season, the number of hours the system is working during the heating season, and last but by no means least, the habits and personal needs of individual people which largely determine the amount of space heat, and hot water which a particular installation will be called upon to provide over any given period.

Most of these factors are infinitely variable, for which reason it is not possible to make an accurate forecast of fuel consumption. By far the most reliable method is by reference to the operating performance, extending over a minimum period of three years or more, of an installation of similar design, size, and ownership to the one under review. Unfortunately very little factual information is available at the present time. Until reliable data on comparative annual costs become available as the result of field research, estimates of future costs must be calculated, using the best available data.

Efficiency

Boiler furnace efficiency is expressed by a fraction giving the percentage of heat from the fuel which is transmitted to the water in the boiler.

Thus if $\frac{80}{100}$ or 80 per cent of the heat released by combustion of the fuel is used in raising the temperature of the boiler water, and the remaining 20 per cent passes up the chimney to waste, without heating the water, we say that the furnace efficiency is 80 per cent.

Boiler efficiency expresses the percentage of heat produced by combustion which actually leaves the boiler for use in the building. Suppose the heat loss from the boiler surfaces to the air in the boiler room amounts to 3 per cent of the heat produced in the boiler, then the boiler efficiency will be equal to

$$\frac{80 \times 97}{100} = 77{\cdot}6 \text{ per cent.}$$

If we further assume that 10 per cent of the heat which leaves the boiler is wasted owing to loss from pipe surfaces in the boiler room and other unheated spaces, the *overall efficiency* of the system will be

$$\frac{77{\cdot}6 \times 90}{100} = 69{\cdot}84 \text{ per cent, say 70 per cent}$$

Overall efficiency of a space-heating system is also affected by excess heat loss at radiator positions on outside walls. For this reason alone, unless extra precautions are taken to prevent excess local heat loss, the overall efficiency of any radiator system will be lower than the boiler efficiency.

In small-bore systems for small dwellings as depicted in Figs. 2A and 2B, very little of the heat generated in the boiler is wasted, and provided care is taken to prevent excess heat loss at radiator positions, the overall efficiency closely approaches the boiler efficiency. 13 mm insulating board lined with aluminium foil to let into the plaster between the radiator and an outside wall will eliminate excessive local heat loss.

Maximum Hourly Fuel Consumption

The maximum hourly fuel consumption is given by the expression

$$\text{Fuel consumed per hour} = \frac{\text{installed heat load in kilowatts} \times 3600}{\text{calorific value of fuel} \times \text{overall efficiency}}$$

Average calorific values of domestic fuels and overall efficiencies are given in Table 12.

Based upon the above the maximum hourly fuel consumption for each fuel, per kilowatt of installed heat load, will be:

Anthracite

$$\text{Fuel per hour} = \frac{1\ \text{kW} \times 3600}{33730\ \text{kJ/kg} \times 0{\cdot}75} = 0{\cdot}1423\ \text{kg h/kW}$$

Hand-fired coke

$$\text{Fuel per hour} = \frac{1\ \text{kW} \times 3600}{29070\ \text{kJ/kg} \times 0{\cdot}55} = 0{\cdot}2252\ \text{kg h/kW}$$

Domestic oil

$$\text{Fuel per hour} = \frac{1\ \text{kW} \times 3600}{36660\ \text{kJ/litre} \times 0{\cdot}75} = 0{\cdot}1309\ \text{litre h/kW}$$

Coal gas

$$\text{Fuel per hour} = \frac{1\ \text{kW} \times 3600}{18630\ \text{kJ/m}^3 \times 0{\cdot}80} = 0{\cdot}2415\ \text{m}^3\ \text{h/kW}$$

Table 12

Average calorific values of fuels commonly used for domestic heating

Fuel	*Overall efficiency (per cent)*	*Heat value (per unit)*
Anthracite (automatically fired in gravity-feed boiler)	75	14500 Btu/lb = 33730 kJ/kg
Coke and similar smokeless fuels hand-fired	55	12500 Btu/lb = 29070 kJ/kg
Domestic oil	75	158000 Btu/gal = 36660 kJ/litre
Gas	80	coal gas: 500 Btu/ft³ = 18630 kJ/m³; natural gas: 1000 Btu/ft³ = 37260 kJ/m³
Electric small-bore boiler unit (off-peak)	90	1 kWh = 3415 × 1·055 = 3600 kJ
Electric floor warming (off-peak)	85	
Electric storage heaters (off-peak)	95	
Electric, on-peak fires, radiators and convectors	100	

Natural gas

$$\text{Fuel per hour} = \frac{1\ \text{kW} \times 3600}{37260\ \text{kJ/m}^3 \times 0{\cdot}80} = 0{\cdot}1208\ \text{m}^3\ \text{h/kW}$$

Electric floor warming

$$\text{Fuel per hour} = \frac{1\ \text{kW} \times 3600}{3600\ \text{kJ} \times 0{\cdot}85} = 1{\cdot}17\ \text{kWh/kW}$$

Electric small-bore heating boiler

$$\text{Fuel per hour} = \frac{1\ \text{kW} \times 3600}{3600\ \text{kJ} \times 0{\cdot}90} = 1{\cdot}111\ \text{kWh/kW}$$

Electric storage heaters

$$\text{Fuel per hour} = \frac{1\ \text{kW} \times 3600}{3600\ \text{kJ} \times 0{\cdot}95} = 1{\cdot}053\ \text{kWh/kW}$$

Electric fires, radiators and convectors

$$\text{Fuel per hour} = \frac{1\ \text{kW} \times 3600}{3600\ \text{kJ} \times 1{\cdot}00} = 1\ \text{kWh/kW}$$

Calculating Seasonal Fuel Consumption

The above figures represent the maximum estimated fuel consumption when the heating system is operating at full design load. For most of the heating season, however, the outside temperature is above the basic design temperature of −1°C, and the hourly fuel consumption will during these periods be something less than the maximum.

To forecast the probable fuel consumption for any period, we must have some guide as to probable demand for heat during that period, and in this connection the *degree day* has been found to produce reasonably accurate results.

The degree day may be defined as the difference between 60°F and the daily mean outside temperature when the daily mean outside temperature is below 60°F. The temperature of an unheated building is usually 5°F higher than the outside temperature, and for this reason the base temperature of 60°F is used.

Table 9(A) Appendix A gives monthly average degree-day figures for the period 1945/1965 in the UK to a base temperature of 60°F, and Table 9(B) gives monthly average degree days in

Ireland over the period 1958/1962 to a base temperature of 60°F.

Table 13 (parts A and B) gives monthly average degree days over the same periods to a temperature base of 15·6°C (60°F),

Table 13

(A) Average for 20 year period 1945-65 in Great Britain. Compiled from data issued by the Gas Council. Base temperature = 15·6°C

Area	*Sep.*	*Oct.*	*Nov.*	*Dec.*	*Jan.*	*Feb.*	*Mar.*	*Apl.*	*May*	*Total*
Thames Valley	54	140	234	318	354	317	280	187	112	1996
South Eastern	74	170	273	358	397	353	313	221	143	2302
Southern	56	129	225	308	347	309	271	183	114	1942
Western	47	113	200	269	309	286	255	181	116	1776
Severn Valley	66	151	247	330	366	328	293	204	131	2116
Midland	74	164	265	348	383	344	305	214	141	2238
Lancashire	81	167	266	354	384	340	305	219	144	2260
North Western	96	179	267	351	385	346	317	231	155	2327
North Eastern	100	186	284	371	400	357	336	241	175	2450
Yorkshire	78	169	270	350	388	343	317	213	140	2268
Eastern	72	161	265	352	388	339	302	207	134	2220
S.E. Scotland	100	183	277	354	388	353	331	241	179	2406
West Scotland	106	197	298	378	413	363	334	239	162	2490
East Scotland	104	192	295	375	405	353	331	237	167	2459

(B) Average for 5 year period 1958-62 in Ireland. Compiled from data issued by the Department of Transport and Power Meteorological Service, Dublin. Base temperature = 15·6°C

Area	Sep.	Oct.	Nov.	Dec.	Jan.	Feb.	Mar.	Apl.	May	Total
Roches Point	54	125	213	275	288	241	258	197	135	1786
Cahirciveen	50	118	198	254	264	218	230	177	121	1630
Rosslare	48	122	210	281	302	253	273	204	141	1834
Kilkenny	87	170	268	348	356	275	293	209	142	2148
Shannon Airport	58	142	242	324	332	259	265	191	121	1934
Birr	81	167	265	349	353	276	290	211	140	2132
Dublin Airport	70	154	247	327	343	273	294	215	146	2069
Mullingar	94	182	282	367	377	297	309	225	150	2283
Claremorris	97	174	273	357	370	290	297	225	156	2239
Clones	97	178	282	370	380	298	310	231	153	2299
Belmullet	65	133	221	296	313	250	267	202	143	1890
Malin Head	73	137	228	304	321	268	286	225	164	2006

derived from Table 9 (parts A and B) Appendix A, using the expression

$$\text{degree day to base of } 60°\text{F} \times 0{\cdot}56 = \text{degree day to base of } 15{\cdot}6°\text{C}.$$

The seasonal fuel consumption of a space-heating system equipped with effective thermostatic control will vary according to the length of the season and the severity of the weather during the season.

An appreciation of the length and severity of the heating season in Scotland, Northern England, Midlands and Southern England can be obtained from a study of Table 14, columns 2 and 3 of which give the season's length, and column 4 the average degree days.

Table 14
Heating season for the four main zones of Great Britain
(Showing the length and severity of the season)

Main zone	*Heating season (inclusive)*	*Number of days in season (length)*	*Total degree days (severity)*
Scotland	Sept. to May	273	2452
Northern England	Oct. to May	243	2237
Midlands	Oct. to April	212	2019
Southern England	Oct. to April	212	1844

For the purpose of calculating domestic heating fuel consumption, it is sufficiently accurate and convenient to divide Great Britain into these four main zones.

Seasonal fuel consumption per 1000 W of installed load can be found by the expression

$$\text{Fuel consumed per 1000W} = \frac{\text{degree days (Table 14, col. 4)} \times 24}{16{\cdot}7} \times C$$

where C is the maximum hourly fuel consumption per 1000W of installed space heating load.

Values of C for the various fuels have already been determined, and to show the use of Table 14 and the method of calculation we shall now estimate the probable seasonal consumption per 1000W of installed heating load, for domestic space heating in the North of England.

Anthracite
Fuel consumption

$$= \frac{2237 \times 24}{16{\cdot}7} \times 0{\cdot}1423 \text{ kg}$$

$$= 3215 \times 0{\cdot}1423 \text{ kg} = 458 \text{ kg season/kW}$$

Coke
Fuel consumption

$$= 3215 \times 0{\cdot}2252 \text{ kg} = 724 \text{ kg season/kW}$$

Domestic oil
Fuel consumption

$$= 3215 \times 0{\cdot}1309 \text{ litre} = 421 \text{ litre season/kW}$$

Coal gas
Fuel consumption

$$= 3215 \times 0{\cdot}2415 \text{ m}^3 = 776 \text{ m}^3 \text{ season/kW}$$

Natural gas
Fuel consumption

$$= 3215 \times 0{\cdot}1208 \text{ m}^3 = 388 \text{ m}^3 \text{ season/kW}$$

Electric floor warming
Fuel consumption

$$= 3215 \times 1{\cdot}17 = 3762 \text{ kWh/kW}$$

Electric small-bore boiler unit
Fuel consumption

$$= 3215 \times 1{\cdot}111 = 3572 \text{ kWh/kW}$$

Electric storage heaters
Fuel consumption

$$= 3215 \times 1{\cdot}053 = 3385 \text{ kWh/kW}$$

Table 15
Estimated fuel consumption for full heating 24 h per day.*

Main zone	*Anthracite* *kg*	*Hand-fired coke* *kg*	*Oil* *litre*	*Coal gas* m^3	*Natural gas* m^3	*Natural floor warming* *kWh*	*Electric storage heaters* *kWh*	*Electric small-bore boiler* *kWh*	*Electric fires, radiators etc.* *kWh*
Scotland	502	796	461	851	426	4124	3711	3915	3525
Northern England	458	724	421	776	388	3762	3385	3572	3215
Midlands	413	655	380	701	351	3394	3055	3223	2901
Southern England	377	598	347	640	320	3100	2790	2944	2650

*Fuel consumption per kilowatt of installed space heating load (radiators, pipes and other heat-losing surfaces) for domestic heating installations with effective thermostatic control, and operating to produce design room temperatures for 24 hours each day.

Appropriate consumption for installations in Ireland may be calculated from Table 13(B).

Electric fires, radiators, convectors, etc.
Fuel consumption

$$= 3215 \times 1{\cdot}000 = 3215 \text{ kWh/kW}$$

Similar calculations for Scotland, the Midlands, Southern England and Ireland, will produce the constants given in Table 15. These values represent the estimated fuel consumption for full heating, 24 hours per day. In most family dwellings, especially where there are children or elderly people, the heating system is required to produce full design temperatures all day, with possibly some reduction in living-room temperatures during the night.

Individual room thermostatic-radiator control enables the system to meet the changing needs of rooms during a 24-hour spell, and will of course lead to considerable reductions in the values given in Table 15. The lounge, dining room and bedrooms may not, in many homes, need to be kept continuously up to design temperature, and the judicious application of thermostatic

Table 16
Multiplying factors for intermittent and reduced temperature operation of small domestic heating installations.
(For use with Table 15).

Operation	*Factor*
All room temperatures reduced by 6°C, from 11 p.m. to 7 a.m.	0·85
All room temperatures reduced by 6°C, from 9 a.m. to 5 p.m.	0·85
System shut down from 11 p.m. to 6 a.m.	electricity, gas and oil, 0·75
System shut down from 11 p.m. to 6 a.m.	solid fuel, 0·80
System shut down from 9 a.m. to 4 p.m.	electricity, gas and oil, 0·75
System shut down from 9 a.m. to 4 p.m.	solid fuel, 0·80
Thermostatic control of individual rooms according to occupation	0·85 to 0·90, depending upon the degree of occupation

radiator control will effect a reduction of 10 to 15 per cent of the values given in Table 15.

Multiplying factors for use in conjunction with Table 15 to allow for intermittent and reduced temperature operation, are given in Table 16.

Estimating Fuel Consumption for Domestic Hot Water

In Chapter 7 we estimated that a normal household of five people, will, on the day of heaviest demand, use 550 litre of hot water. This provided a basis for estimating the capacity of the cold-water storage tank, for which purpose maximum hot-water usage had to be assumed.

Fuel consumption will of course depend upon the amount of hot water used during the year (52 weeks), and also the heat loss from pipes and cylinder during that period.

Hot-water usage varies according to the number of people in the house, their personal requirements and habits, and also according to the location of the dwelling. A family in a dirty smoky town will use more hot water than a family of similar size and habits living in the country.

The average weekly hot-water consumptions for a modern three-bedroom semi-detached house, as estimated by the author are given in Table 17.

Table 17

Average weekly hot water consumed by families living in a three-bedroom house of approximately 93 m^2 area.

Number in family	*Weekly hot-water consumption (litre)*	
	House occupied for 24 hours on each day of week	*House closed during daytime. Monday to Friday inclusive*
5	1818	1455
4	1546	1182
3	1273	1000
2	909	727

Heat is continually being emitted from the hot-water cylinder and primary pipes, and in addition heat is wasted owing to water cooling in the secondary dead leg after water has been drawn from the system. Heat loss from pipes and cylinder will remain fairly constant throughout the year, while the loss due to cooling will vary according to the number of times the taps are used. It is reasonable to assume that the frequency with which taps are used will be largely influenced by the number of people living in the house. The habits of the residents will also have a marked effect upon heat waste from this cause. Table 18 has been compiled in order that some allowance may be made for this heat loss when estimating the fuel consumption for domestic hot-water systems, and represents maximum usage of hot water by the family.

Table 18
Heat loss due to water cooling in secondary pipes of hot-water installation in three-bedroom semi-detached house (approx. floor area 93m^2).

Number in family	*Daily heat loss (kJ/24 h)*	
	House occupied for 24 hours on each day of week	*House closed during daytime. Monday to Friday inclusive*
5	12670	7596
4	10760	6436
3	8860	5275
2	6330	3798

In the compilation of Table 18 the cooling of the secondary piping system serving a family of five is assumed to take place, on average, twenty times each day. The secondary pipes in the average three-bedroom house contain approximately 3·4 litre of water, which, if allowed to cool from the storage temperature of 65·6 to 21·1°C twenty times per day, will waste

$$3{\cdot}4 \times 20 \times (65{\cdot}6 - 21{\cdot}1) \times 4{\cdot}187 = 12670 \text{ kJ/day.}$$

Using Tables 15–18, a reasonably accurate forecast of fuel consumption for any type of domestic space heating, and hot-water supply system can be made, with the very minimum of calculation.

As an example, we shall estimate the probable annual fuel consumption of the installation which formed the subject of the earlier chapters of this book.

$$\text{Space heating load} = 11{\cdot}29 \text{ kW}.$$

Assuming the system to be installed in the north of England, the annual fuel consumption for full twenty-four hour heating, from Table 15, will be:

Anthracite $\dfrac{458 \text{ kg} \times 11{\cdot}29}{1000} = 5{\cdot}171$ tonnes

Coke $\dfrac{724 \times 11{\cdot}29}{1000} = 8{\cdot}174$ tonnes

Oil $421 \times 11{\cdot}29 = 4753$ litre

Coal gas $701 \times 11{\cdot}29 = 7914 \text{ m}^3$

Natural gas $351 \times 11{\cdot}29 = 3963 \text{ m}^3$

Electric small-bore boiler unit

$$3572 \times 11{\cdot}29 = 40330 \text{ kWh}$$

The total space-heating load for small-bore hot-water heating of 11·29 kW includes 1·02 kW for mains and other surface losses which are not present with direct electric space-heating systems, and for these the total space heat load is reduced to $11{\cdot}29 - 1{\cdot}02 = 10{\cdot}27$ kW.

Electric floor warming $3762 \times 10{\cdot}27 = 38640$ kWh

Electric storage heaters $3385 \times 10{\cdot}27 = 34760$ kWh

Electric fires, radiators and convectors
$3215 \times 10{\cdot}27 = 33020$ kWh

Hot-water supply

Assume a family of three in full daily occupation. From Table 17, weekly heat for hot water

$$= 1273 \text{ litre} \times (65{\cdot}6°\text{C} - 10°\text{C}) \times 4{\cdot}187 = 296300 \text{ kJ}.$$

From Table 18, weekly heat loss due to water cooling in secondary pipes

$$= 8860 \text{ kJ/day} \times 7 = 62020 \text{ kJ}.$$

Weekly heat loss from insulated cylinder and primary pipe surfaces

$$= 0{\cdot}248 \text{ kW} \times 168 \text{ hours} \times 3600 = 149800 \text{ kJ}$$

Summary of hot-water heat requirements:

Heat to produce hot water	=	296300 kJ
Heat to replace cooling	=	62020 kJ
Heat loss from surfaces	=	149800 kJ
Total weekly heat for HWS	=	508120 kJ

It should be noted at this stage that, where electricity is the heating medium, an immersion heater will be employed, and primary pipework will be unnecessary. Consequently the weekly heat requirements for hot water are reduced to 446100 kJ.

The annual consumption of each fuel to produce the hot water for the household can now be calculated (see Table 12 for efficiencies and calorific values).

Anthracite $\dfrac{508120 \times 52 \text{ weeks}}{33730 \times 0{\cdot}75 \times 1000} = 1{\cdot}044$ tonnes

Coke $\dfrac{508120 \times 52}{29070 \times 0{\cdot}55 \times 1000} = 1{\cdot}652$ tonnes

Oil $\dfrac{508120 \times 52}{36660 \times 0{\cdot}75} = 961$ litres

Coal gas $\dfrac{508120 \times 52}{18630 \times 0{\cdot}80} = 1773 \text{ m}^3$

Natural gas $\dfrac{508120 \times 52}{37260 \times 0{\cdot}80} = 887 \text{ m}^3$

Electricity $\dfrac{446100 \times 52}{3600^*} = 6443$ kWh

The total annual fuel consumption will be:

Anthracite: Heating 5·171 t + HWS 1·044 t = 6·215 tonnes

*1kWh = 3600 kJ.

Coke: Heating 8·174 t + HWS 1·652 t = 9·826 tonnes

Oil: Heating 4753 litres + HWS 961 litres = 5714 litres

Coal gas: Heating 7914 m^3 + HWS 1773 m^3 = 9687 m^3

Natural gas: Heating 3963 m^3 + HWS 887 m^3 = 4850 m^3

Electric small-bore boiler unit
Heating 40330 kWh + HWS 6443 kWh = 46773 kWh

Electric floor warming
Heating 38640 kWh + HWS 6443 kWh = 45083 kWh

Electric storage heaters
Heating 34760 kWh + HWS 6443 kWh = 41203 kWh

Electric fires, radiators and convectors
Heating 33020 kWh + HWS 6443 kWh = 39463 kWh

We are now in a position to compare the estimated annual cost of various fuels, and the actual costing in this and the next (and final) chapter, is expressed in decimal currency. The decimal prices per SI unit of measure quoted below, are related to the UK coinage values per Imperial unit of measure, current at the time of writing.

Anthracite

6·215 t at £20·18 per t = £125·40

Coke

9·826 t at £15·36 per t = £151·00

Oil

$$\frac{\text{5714 litres at 1·925 p/litre}}{100} = £110·00$$

Coal gas

The present two-part tariff makes a standing charge of £8·67

*See Appendix B for information on booklets dealing with decimal currency.

(£8 13s 4d) per annum plus a commodity charge of 1s 6d per therm. Under the new decimal/metric system of currency and mensuration, gas will be measured in cubic metres and charged on cost per megajoule (MJ).

In decimal currency 1 p (new penny) = 2·4d, and the cost per therm

$$= \frac{1s.\ 6d}{2{\cdot}4} = \frac{18d}{2{\cdot}4}$$

As 1 therm = 105·5 MJ the cost per MJ, will be:

$$\frac{18}{2{\cdot}4 \times 105{\cdot}5} = 0{\cdot}071 \text{ p/MJ}$$

and the cost of 9687 m^3 of Town gas

$$= £8{\cdot}67 + \frac{9687 \text{ m}^3 \times 18{\cdot}63 \text{ MJ/m}^3 \times 0{\cdot}071 \text{ p}}{100}$$

$$= £8{\cdot}67 + £128{\cdot}10 = £136{\cdot}77.$$

Natural gas

The tariff to be paid by users of natural gas includes an annual standing charge of £15·60 (£15 12s. 0d) plus a commodity charge of 1s. 3d per therm, which is equivalent in decimal/metric terms to a commodity charge of

$$\frac{15d}{2{\cdot}4 \times 105{\cdot}5} = 0{\cdot}059 \text{ p/MJ}$$

and the cost of 4850 m^3 of natural gas

$$= £15{\cdot}60 + \frac{4850 \text{ m}^3 \times 37{\cdot}26 \text{ MJ/m}^3 \times 0{\cdot}059 \text{ p}}{100}$$

$$= £15{\cdot}60 + £106{\cdot}90 = £122{\cdot}50.$$

Electric small-bore heating boiler unit (off-peak)

$$\frac{46773 \text{ kWh} \times 0{\cdot}3125 \text{ p*}}{100} = £146{\cdot}20$$

Electric floor warming (off-peak)

$$\frac{45083 \text{ kWh} \times 0{\cdot}3542 \text{ p†}}{100} = £159{\cdot}70$$

*0·75d = 0·3125 p.
†0·85d = 0·3542 p.

Electric storage heaters (off-peak)

$$\frac{41203 \text{ kWh} \times 0{\cdot}3542 \text{ p}}{100} = £145{\cdot}90$$

Electric fires, radiators and convectors (white meter tariff)

The 'white meter' tariff allows the consumer to use electricity for all purposes at the off-peak rate of 0·3125 p (0·75d) per unit during the 8 hours, 11 p.m. to 7 a.m., and at 0·8335 p (2d) per unit between 7 a.m. and 11 p.m.

For an installation of oil-filled radiators operating 24 h/day on this tariff the probable annual cost of electricity will be:

$$\text{Day period:} \quad \frac{39463 \text{ kWh} \times 2 \times 0{\cdot}8335 \text{ p}}{3 \times 100} = £219{\cdot}30$$

$$\text{Night period:} \quad \frac{39463 \text{ kWh} \times 1 \times 0{\cdot}3125 \text{ p}}{3 \times 100} = £\ 41{\cdot}11$$

$$\text{Total} = £260{\cdot}41$$

The price of fuels varies in different parts of the country, and solid-fuel prices are generally lower by approximately 10 per cent during the summer months. The price per litre of domestic fuel oil varies with the amount of the delivery. From 182 up to 905 litre is charged at one price, 909 up to 2268 litre is charged at a lower rate, and above this at a still lower rate. In small domestic properties, however, space for a tank larger than 1250 litre is rarely available.

Electricity, if used only during restricted hours, for thermal storage forms of space, and water heating may be purchased at rates, which are competitive with other fuels.

The above estimates represent the annual cost of fuel for a thermostatically controlled system, in the North of England, operated to give full heating at design temperatures for twenty four hours, each day of the heating season of 243 days, plus hot water supply for 365 days.

Under certain conditions, an installation in a house identical in every way to the one we are considering will show annual fuel costs which are very much lower than the above estimates. The conditions which will produce these lower annual fuel costs are listed below, and in a specific case one or a combination of these conditions may contribute to the reduced cost.

1. The fuel may be cheaper to buy in the particular area.
2. The house may be in the Midlands, or in the South of England where the heating season is both shorter and milder.
3. The performance of the installation is purposely restricted by the owner to limit fuel costs at the expense of comfort.
4. Intermittent operation is practised, because the house is occupied intermittently.
5. Room temperatures are depressed during selected periods of the day or night, or during certain days of the week, when the rooms are unoccupied.
6. Radiators in certain rooms may be permanently shut off, except in severe weather. Bedrooms occupied by the middle-aged and young are typical examples of this selective operation.
7. During the summer, the house may be unoccupied for certain periods, when both the hot-water and heating systems will be shut down.
8. During the summer, the owner may decide not to operate the boiler, and to rely upon an electric immersion heater for hot water.

Quite obviously the effects of these alternatives to full operation, and of the infinitely variable habits and demands of householders, cannot be precisely assessed, but careful investigation into claims of 'lower than expected annual fuel consumption' has invariably shown the discrepancy to be attributable to one or more of the practices enumerated above.

In Table 16, multiplying factors are given for intermittent and reduced temperature operations commonly practised, and these factors used in conjunction with Table 15 cover a wide range of domestic heating practice.

Consider a house in the north of England where the room temperatures are reduced by 6°C from 11 p.m. to 7 a.m., and where the radiators are controlled individually by room thermostats.

Applying Tables 15 and 16, the probable fuel consumption for space heating will be found by multiplying the annual consumption obtained from Table 15 by the constants given in Table 16, the expression being:

Fuel consumption for continuous heating × 0·85 × 0·85 = fuel consumption for continuous heating × 0·72.

For the house under review the revised comparative annual

fuel cost estimates will be as follows:

Anthracite

Heating $(5{\cdot}171 \times 0{\cdot}72)$ + HWS $1{\cdot}044$ t

$= 3{\cdot}723 + 1{\cdot}044 = 4{\cdot}767$ t

Cost $= 4{\cdot}767$ t at £20·18/t = £96·18

Coke

Heating $(8{\cdot}174 \times 0{\cdot}72)$ + HWS $1{\cdot}652$ t

$= 5{\cdot}885 + 1{\cdot}652 = 7{\cdot}537$ t

Cost $= 7{\cdot}537$ t at £15·16/t = £115·80

Oil

Heating (4753 litres $\times$ 0·72) + HWS 961 litre

$= 3422 + 961 = 4383$ litre

Cost = 4383 litres at 1·925 p/litre = £84·37

Coal gas

Heating (7914 m^3 $\times$ 0·72) + HWS 1773 m^3

$= 5698 + 1773 = 7471\ m^3$

$$\text{Cost} = \frac{7471\ m^3 \times 18{\cdot}63\ MJ/m^3 \times 0{\cdot}071\ p}{100} + £8{\cdot}67$$

$= £98{\cdot}81 + £8{\cdot}67 = £107{\cdot}48$

Natural gas

Heating (3963 m^3 $\times$ 0·72) + HWS 887 m^3

$= 2853 + 887 = 3740\ m^3$

$$\text{Cost} = \frac{3740\ m^3 \times 37{\cdot}26\ MJ/m^3 \times 0{\cdot}059\ p}{100} + £15{\cdot}60$$

$= £82{\cdot}22 + £15{\cdot}60 = £97{\cdot}82$

Electric small-bore boiler unit

Heating (40330 kWh × 0·72) + HWS 6443 kWh

$$= 29028 + 6443 = 35471 \text{ kWh}$$

$$\text{Cost} = \frac{35471 \text{ kWh} \times 0{\cdot}3125 \text{ p}}{100} = £110{\cdot}90$$

Electric storage heaters

Heating (34760 × 0·72) + HWS 6443 kWh

$$= 25027 + 6443 = 31470 \text{ kWh}$$

$$\text{Cost} = \frac{31470 \times 0{\cdot}3542 \text{ p}}{100} = £111{\cdot}40$$

Building heat losses for the example in this book have rightly been based on normal U values.

Structural insulation, applied to the standards recommended in Appendix F, will reduce the building heat loss by 27 per cent, and so produce a most useful reduction in fuel consumption.

An approximate analysis of the reduction follows:

Insulated roof will reduce the total loss by	13 per cent
Insulated external walls reduce total loss by	9 per cent
Insulated solid ground floor reduces total loss by	5 per cent
Total reduction	27 per cent

9

Comparison of Annual Costs using Different Fuels

Fuel is but one item in the build-up of annual cost. The next largest item is the capital charge, or annual charge for depreciation. A write-off period of ten years is often adopted for domestic heating equipment, which means that an annual charge for one-tenth of the capital cost of an installation must be allowed. In cases where the owner chooses to pay for the installation out of income on deferred terms he will include the amount of his annual capital repayment plus interest, thus in lieu of a depreciation charge. Alternatively, the project may be financed through a building society, in which case repayment of up to 90 per cent of the capital cost plus interest may be spread over a much longer period than ten years. All the principal fuel interests can arrange credit loan facilities which enable the owner to spread the whole of the capital cost over periods up to ten years.

Other items of annual cost must cover insurance; service charges; meter charges; spares and replacements; flue cleaning; electricity for pumps, fans, damper motors and controls; cost of redecoration caused by boilers fixed in kitchen or other room.

Table 19 compares the capital cost build-up for the small-bore space heating and hot-water supply installation referred to in Chapter 2 using the various available fuels.

A comparison of the annual costs for this installation, when operating continuously under thermostatic plant control to meet design conditions, is given in Table 20.

Table 21 compares annual costs when running at depressed night room temperatures, and having the benefit of individual thermostatic room control.

Table 19

Comparison of capital cost*

Item of capital cost	*Anthracite automatic boiler S.B.S.†* £	*Coke hand-fired boiler S.B.S* £	*Domestic oil boiler S.B.S.* £	*Natural-gas boiler S.B.S.* £	*Electric small-bore boiler unit S.B.S.* £	*Electric storage heaters* £
Engineering works	470	420	520	470	470	360
Builders work	55	50	50	50	20	—
Fuel storage facilities	20	20	35	—	—	—
Total cost	545	490	605	520	490	360
Annual charge based upon a 10 year write-off	54·50	49·00	60·50	52·00	49·00	36·00

*Small-bore low-pressure hot water, pump-operated space heating and hot-water supply installation for a small semi-detached house. The work carried out by fully qualified trades unionist craftsmen (for recognised central-heating contractors) during normal working hours, and to a good engineering specification complying with the relevant British Standard specifications, and the codes of practice laid down by the IHVE, and the HVCA, the IEE and ECA whichever is appropriate to the type of space heating and hot-water supply installed.
† Small-bore system.

Table 20

Comparison of annual operating costs*

Item of annual cost	*Anthracite automatic boiler S.B.S.*†	*Coke hand-fired boiler S.B.S.*	*Domestic oil boiler S.B.S.*	*Natural-gas boiler S.B.S.*	*Electric small-bore boiler unit S.B.S.*	*Electric storage heaters*
	£	£	£	£	£	£
Capital charge	54·50	49·00	60·50	52·00	49·00	36·00
Fuel	125·40	151·00	110·00	122·50	146·20	145·90
Service charge	—	—	12·00	05·00	—	—
Insurance (13 p per cent)	0·61	0·55	0·68	0·61	0·61	0·47
Damage to decorations	6·50	6·50	4·80	3·20	3·20	—
Spares and replacements	2·25	0·75	4·50	3·00	2·00	1·50
Standing or meter charges	—	—	—	—	0·63	0·63
Flue cleaning	0·75	0·75	0·75	0·75	—	—
Electricity for pumps etc.	2·25	1·50	3·00	1·50	1·0	—
Annual total	192·26	210·05	196·23	188·56	202·64	184·50

* Domestic space heating and hot-water supply installation in a small semi-detached 3-bedroom house in the North of England. The house is in continuous occupation 365 days a year. Heating season October to May inclusive, 243 days, (total average degree days 2237 to base of 15·6°C). The installation operates under full thermostatic plant control to provide full design temperatures in all rooms for 24 hours each day, and continuous hot water at a temperature of 60°C at the taps.

† Low-pressure small-bore hot-water space heating.

Table 21

Comparison of operating costs*

Item of capital cost	*Anthracite automatic boiler S.B.S.†*	*Coke hand-fired boiler S.B.S.*	*Domestic oil boiler S.B.S.*	*Natural-gas boiler S.B.S.*	*Electric small-bore boiler unit S.B.S.*	*Electric storage heaters*
	£	£	£	£	£	£
Capital charge	54·50	49·00	60·50	52·00	49·00	36·00
Fuel	96·18	115·80	84·37	97·38	110·90	111·40
Other items as Table 20	12·86	10·05	25·41	14·06	07·44	2·60
Total annual cost	163·54	174·85	170·28	163·44	167·34	150·00

* Domestic space heating and hot-water supply installation as described for Table 20, but operating at reduced night temperature (6°C depression) from 11 p.m. to 7 a.m., and having individual thermostatic radiator control in addition to the overall plant control provided for in Table 20.
The electric storage heater installation is controlled by an 'inside-outside temperature' system in addition to the time-switch.
† Low-pressure small-bore hot-water space heating.

The Effect of Central Heating on Rateable Value

When considering the provision of any form of fixed space heating, the prospective owner should be informed of the possible increase in rateable value of his property.

The basis of valuation for rating is the annual rent which a hypothetical tenant would be expected to pay the owner. If the owner installs a fixed form of space heating, namely, radiators, convectors, pipes, hot-air ducts, permanently-fixed central heaters or boilers, or electric floor warming, he would expect some return from this investment. The rating valuation officer will make a proper assessment of this, and add an appropriate amount to the rateable value of the property.

When comparing annual costs it is therefore necessary in the case of coal, coke, anthracite, oil- and gas-fired central-heating pipe and radiator systems, central air-heating systems and electric floor warming, to include an additional item of annual cost to cover the increase in annual rates payable to the local authority.

Movable, or portable forms of gas, oil and electric heaters are regarded as 'goods and chattels', and as such do not affect the rateable value of the property. These include certain portable gas heaters, oil radiators and convectors, electric fires, convectors, oil-filled radiators and storage heaters.

An assessment of this nature should refer only to that part of the capital cost of the installation which has been due to the provision of space heating as distinct from domestic hot-water heating, because the last mentioned is a common amenity, the value of which will be reflected in the normal rating assessment of every modern property. Referring to Tables 20 and 21 therefore, the total amounts given in Columns 2–6 should, where the local authority requires it, be increased to cover this additional rating charge.

Appendix A Domestic Heating Design Tables in Imperial Units

Table 1

Heat transmittance coefficients (U: Btu/ft^2 h °F)

Orientation	*Exposure to wind*					
S	*Sheltered*	*Normal*	*Severe*			
W, SW, SE		*Sheltered*	*Normal*	*Severe*		
NW			*Sheltered*	*Normal*	*Severe*	
N, NE, E			*Sheltered*	*Normal*		*Severe*
Brick walls						
4½ in. brick, plastered one side	0·46	0·49	0·53	0·57	0·61	0·65
9 in. brick, plastered one side	0·36	0·38	0·41	0·43	0·45	0·48
11 in. cavity brick, plastered one side (unventilated)	0·27	0·28	0·29	0·30	0·31	0·32
11 in. cavity brick, plastered one side (ventilated)	0·30	0·31	0·33	0·34	0·36	0·37
Windows						
Single glazing	0·70	0·79	0·88	1·00	1·14	1·30
Double glazing	0·41	0·44	0·47	0·50	0·53	0·56

Ground floors	*U* for inside–outside temperature difference
(1) Solid concrete in contact with earth	0·20
(2) As (1) with fitted carpet and underlay	0·16
(3) Ventilated wood floor on joists, air bricks on one side, bare boards	0·30
(4) As (3) with fitted carpet and underlay	0·22
(5) As (3) with air bricks on more than one side	0·40
(6) As (5) with fitted carpet and underlay	0·27

Intermediate floors	Heat flow downwards	Heat flow upwards
(1) Wood on joists, plaster ceiling	0·22	0·29
(2) As (1) with fitted carpet and underlay	0·17	0·21

	Sheltered	Normal	Severe
Roofs (flat)			
(1) 4 in. concrete, asphalted, with plaster ceiling	0·61	0·67	0·73
(2) As (1) with 1 in. glass fibre or mineral wool board under asphalt	0·18	0·20	0·21
Roofs (pitched)			
(1) Tiles on battens, plaster ceiling; roof space	0·50	0·56	0·64
(2) As (1) with 1 in. glass fibre, or mineral wool insulation between or over ceiling joists	0·17	0·18	0·20

Special Note: For complete lists of *U* values see publications by The Institution of Heating and Ventilating Engineers, 49 Cadogan Square, London SW1. The Structural Insulation Association, 32 Queen Anne Street, London W1 and The British Standards Institution, 2 Park Street, London W1.

The above factors are quoted mainly from the IHVE Guide 1965. For SI values see Table 2, Chapter 1.

Table 2

Heat emission (Btu/h lineal ft) from uninsulated unpainted copper pipes (BS 659) run singly adjacent to walls as in domestic small-bore central heating and installed in heated and unheated spaces (Table 4 footnotes apply here also)

Nominal bore of pipe in	*Space temperature*						
	70°F	65°F	60°F	55°F	50°F	40°F	30°F
(1)	(2)	(3)	(4)	(5)	(6)	(7)	(8)
$\frac{1}{4}$	18	20	21	22	23	25	27
$\frac{3}{8}$	22	24	25	26	27	30	32
$\frac{1}{2}$	27	29	31	33	35	46	49
$\frac{3}{4}$	36	39	41	43	46	62	66
1	45	47	50	54	56	77	82
$1\frac{1}{4}$	54	56	60	64	66	92	98

For SI values see Table 4, Chapter 2.

Table 3

Maximum recommended heat flow and water-flow rates of copper pipes (BS 659) used for small-bore central-heating systems in houses. Water velocity = 3 ft/s; temperature drop in circuit = 20°F

Nominal bore of pipe in	*Maximum heat flow rate Btu/h*	*Maximum water flow rate lb/h*
$\frac{1}{4}$	5400	270
$\frac{3}{8}$	11400	570
$\frac{1}{2}$	19540	977
$\frac{3}{4}$	42600	2130
1	74000	3700
$1\frac{1}{4}$	113600	5680

For SI values see Table 5, Chapter 2.

Table 4

Heat emission (Btu/h ft²) from steel radiators used for small-bore systems operating at 180°F flow and 160°F return. Radiator mean temperature = 170°F

Radiator pattern	*Space temperature*			
	70°F	65°F	60°F	55°F
(1)	(2)	(3)	(4)	(5)
Single-panel	192	205	217	230
Double-panel	165	176	186	198
Treble-panel	146	156	165	176
Two-column	158	169	178	190
Three-column	165	176	186	198
Four-column	160	170	180	190
6 in. skirting heaters*	115	125	135	148

* Outputs for skirting heaters are given in Btu/u per lineal ft.
For SI values see Table 6, Chapter 2.

Table 5

Average heat emission to room from radiator connections when radiator is above and adjacent to the concealed horizontal mains. The heat emission from long final sub-circuits should be estimated separately using Table 2, Appendix A.

Nominal bore of pipe	*Space temperature*			
	70°F	65°F	60°F	55°F
in	Btu/h	Btu/h	Btu/h	Btu/h
(1)	(2)	(3)	(4)	(5)
$\frac{1}{4}$	67	75	78	82
$\frac{3}{8}$	82	90	93	97
$\frac{1}{2}$	100	108	116	123
$\frac{3}{4}$	134	146	152	160

For SI values see Table 7, Chapter 2.

Table 6

Minimum window-sill heights for wall-fixed radiators in dwellings

	Column radiators catalogue height (in)					*Panel radiators catalogue height (in)*			
	14	18	24	30	36	11	18	24	30
(A) Top of the radiator to top of the window sill (inches)	$2\frac{1}{2}$	$2\frac{1}{2}$	$2\frac{1}{2}$	$2\frac{1}{2}$	$2\frac{1}{2}$	$2\frac{1}{2}$	$2\frac{1}{2}$	$2\frac{1}{2}$	$2\frac{1}{2}$
(B) Bottom of radiator to finished floor (inches)	$5\frac{1}{2}$	$5\frac{1}{2}$	$5\frac{1}{2}$	$5\frac{1}{2}$	$5\frac{1}{2}$	$5\frac{1}{2}$	$5\frac{1}{2}$	$5\frac{1}{2}$	$5\frac{1}{2}$
(C) Top of window sill to finished floor (minimum inches)	19	23	29	35	41	19	26	32	38

For SI values see Table 8, Chapter 3.
Note: This table to be read in conjunction with Fig. 4.

Table 7

Recommended minimum length (L) of radiator (in inches) when connected to a one-pipe ½ in. small-bore circuit as shown in Fig. 5, and having ½ in connections. For intermediate flow rates (W_1) and other values of W_2, mean values of L may be computed from columns 2-9 as explained in Chapter 5.

W_1 *Radiator flow rate lb/h*	*Values of W_2 in lb/h*							
	850	800	750	700	650	600	550	500
(1)	(2)	(3)	(4)	(5)	(6)	(7)	(8)	(9)
100	6	6	7	8	9	10	11	14
125	8	9	10	11	13	14	17	20
150	11	12	13	15	17	20	23	27
175	14	15	17	20	22	25	30	36
200	17	19	22	25	28	32	37	45
225	21	24	27	30	34	39	46	55
250	24	27	30	34	39	44	52	63
275	30	34	38	42	48	55	64	77
300	35	39	43	49	55	63	74	90
325	40	45	50	57	65	74	87	104
350	46	52	58	65	74	84	99	120
375	52	58	65	74	83	95	112	134
400	58	64	72	82	94	106	124	149

For SI values see Table 10, Chapter 5.

Table 8

Hot-water supply heat values to be added to space heat requirements when sizing boilers for combined systems in dwellings

Storage capacity of indirect copper cylinder BS 1566 *gal*	*Heat allowance for boiler sizing* *Btu/h*
22–27	8000
32	10000
37	12000
47	16000

For SI values see Table 11, Chapter 6.

In addition to the normal hot-water requirements, the values in column 2 allow for heat emission from the lagged cylinder or the unlagged cylinder in closed airing cupboard. Also the heat emission from insulated or enclosed primary pipework not exceeding 25 ft in length.

Table 9

(A) Average for 20 year period 1945-65 in Great Britain. (Compiled from data issued by the Gas Council) Base temperature = 60°F

Area	*Sep.*	*Oct.*	*Nov.*	*Dec.*	*Jan.*	*Feb.*	*Mar.*	*Apl.*	*May*	*Total*
Thames Valley	97	252	424	572	636	569	504	335	201	3590
South Eastern	133	306	490	644	713	635	561	397	254	4133
Southern	100	232	404	553	622	552	486	326	204	3479
Western	84	201	357	480	552	511	455	323	207	3170
Severn Valley	118	270	441	588	653	585	524	364	234	3777
Midland	132	292	473	622	684	615	554	383	251	4006
Lancashire	145	299	475	633	686	607	545	391	257	4038
North Western	172	319	476	626	687	618	566	413	277	4154
North Eastern	179	333	507	662	714	638	600	431	313	4377
Yorkshire	140	302	482	642	693	612	566	380	250	4067
Eastern	128	288	473	629	692	605	539	369	240	3963
S.E. Scotland	178	327	495	633	693	630	591	430	319	4296
West Scotland	190	351	532	675	737	648	597	427	289	4446
East Scotland	185	342	526	670	724	630	591	424	299	4391

For SI values see Table 13, Chapter 8.

(B) Average for 5 year period 1958-62 in Ireland. (Compiled from data issued by the Department of Transport and Power, Meteorological Service, Dublin). Base temperature = 60°F

Roches Point	96	223	381	494	514	430	460	352	241	3191
Cahirciveen	89	211	353	453	471	389	410	316	216	2908
Rosslare	85	218	376	501	539	452	487	365	252	3275
Kilkenny	156	304	479	622	636	491	523	374	253	3838
Shannon Airport	104	253	433	568	592	462	473	341	216	3442
Birr	144	299	473	617	631	493	517	377	250	3801
Dublin Airport	125	275	441	584	613	487	525	384	260	3694
Mullingar	168	324	503	656	674	531	552	402	268	4078
Claremorris	173	310	488	638	661	517	531	401	278	3997
Clones	172	318	503	660	678	532	554	412	274	4103
Belmullet	116	238	395	528	559	447	477	360	255	3375
Malin Head	130	244	408	541	574	479	511	402	293	3582

For SI values see Table 13, Chapter 8

Table 10

Heating season for the four main zones of Great Britain. (Showing length and severity of the season)

Main zone	*Heating season (inclusive)*	*Number of days in season (length)*	*Average total degree days (severity)*
Scotland	Sept. to May	273	4378
Northern England	Oct. to May	243	4000
Midlands	Oct. to April	212	3609
Southern England	Oct. to April	212	3305

For. SI values see Table 14, Chapter 8.

Table 11

Estimated fuel consumption for full heating, 24 hours per day.*

Main zone	*Anthracite automatic boiler*	*Hand-fired coke boiler*	*Domestic oil boiler*	*Coal-gas boiler*	*Natural-gas boiler*	*Floor warming*	*Electric storage heaters*	*Heating electric small-bore boiler*	*Fires radiators etc.*
	lb	*lb*	*gal*	*ft^3*	*ft^3*	*kWh*	*kWh*	*kWh*	*kWh*
Scotland	324	514	30	8829	4415	1209	1087	1148	1033
Northern England	295	467	27	8052	4026	1103	992	1047	942
Midlands	267	423	25	7240	3620	995	895	945	850
Southern England	243	386	22	6639	3320	909	818	863	777

*Fuel consumption per 1000 Btu/h of installed space heating load (radiators, pipes and other heat losing surfaces) for small domestic installations with effective thermostatic control, and operating to produce design room temperatures for 24 hours each day. Appropriate consumptions for installations in Ireland may be calculated from Table 9(B) Appendix A.
For SI values see Table 15, Chapter 8.

Table 12

Average weekly hot water consumed by families living in a three-bedroom house of approximately 1000 ft^2 area.

Number in family	*Weekly hot-water consumption (gal.)*	
	House occupied for 24 h on each day of the week	*House closed during daytime. Monday to Friday inclusive*
5	400	320
4	340	260
3	280	220
2	200	160

For SI values see Table 17, Chapter 8.

Table 13

Heat loss due to water cooling in secondary pipes of hot-water supply installation in three-bedroom semi-detached house (approximate floor area 1000 ft^2)

Number in family	*Daily heat loss (Btu/ 24h)*	
	House occupied for 24 h on each day of the week	*House closed during daytime. Monday to Friday inclusive.*
5	12000	7200
4	10200	6100
3	8400	5000
2	6000	3600

For SI values see Table 18, Chapter 8.

Table 14

Resistance of copper fittings, etc., normally used for small-bore pumped hot-water heating systems: Equivalent length of straight pipe in feet; comp. = compression fitting; cap. = capillary fitting.

Description of local resistances	*Nominal pipe bore (in)*					
	$\frac{1}{4}$	$\frac{3}{8}$	$\frac{1}{2}$	$\frac{3}{4}$	1	$1\frac{1}{4}$
Regulating valve fully open			24	28	40	
,, ,, $\frac{3}{4}$,,			40	60	70	
,, ,, $\frac{1}{2}$,,			90	150	180	
,, ,, $\frac{1}{4}$,,			500	950	1000	
Gate valve (full way)	0·37	0·61	0·82	1·4	2·0	2·6
Globe valve or tap	6·7	11·0	14·70	24·0	33·5	43·0
Radiator valve (angle)	3·4	5·6	7·5	12·5	18·0	23·0
Elbow comp.	0·82	1·40	1·8	2·9	4·0	4·9
Elbow cap.	0·52	0·85	1·2	1·9	2·7	3·5
Bend comp. or cap.	0·38	0·62	0·83	1·3	1·9	2·4
Through tee comp. or cap.	0·07	0·12	0·16	0·26	0·4	0·5
Tee: comp. or cap.						
Flow into branch	0·9	1·5	2·0	3·2	4·5	5·9
Flow out of branch	1·0	1·7	2·2	3·7	5·3	6·8
Reducer tee branch comp. or cap.	1·1	1·8	2·4	4·0	5·7	7·3
Pitcher tee comp.						
Flow into branch	0·58	0·95	1·3	2·2	3·2	4·2
Flow out of branch	0·65	1·0	1·4	2·5	3·2	4·0
Pitcher tee cap.						
Flow into branch	0·72	1·2	1·5	2·4	3·4	4·4
Flow out of branch	0·72	1·2	1·6	2·5	3·4	3·9
Twin elbow comp.						
Flow into branch	0·44	0·71	1·6	1·9	2·3	2·9
Flow out of branch	0·40	0·65	1·45	1·9	2·1	2·7
Twin elbow cap.						
Flow into branch	0·62	0·98	1·3	2·0	2·7	3·3
Flow out of branch	0·62	1·0	1·4	2·2	3·1	4·0
Return bend (long sweep) comp. or cap.	0·49	0·8	1·1	1·65	2·2	2·7
Reducing coupling comp. or cap.	0·20	0·33	0·45	0·75	1·0	1·4
Outlet from pump, tank or cylinder			2·2	3·7	5·3	7·5
Hot-water boiler			7·0	9·0	12·0	16·0
Radiator	3·0	5·0	8·0	13·0	18·0	—

For SI values see Table 9, Chapter 4.

Appendix B
Metric Conversion Factors
Drawing Scales
Notes on Decimal Currency

Schedule B1

Schedule of SI units and conversions appropriate to the text of this book

Description	*Imperial unit*	*SI unit*	*Conversion*
Length	ft	metre (m)	1 ft = 0·3048 m 1 m = 3·281 ft
	in	milli-metre (mm)	1 in = 25·4 mm
Mass	lb	kilo-gramme (kg)	1 lb = 0·4536 kg 1 kg = 2·205 lb
Time	h, min, s	h, s	
Area	ft^2	m^2	1 ft^2 = 0·0929 m^2 1 m^2 = 10·764 ft^2
Volume	ft^3	m^3	1 ft^3 = 0·02832 m^3 1 m^3 = 35·315 ft^3
Temperature	°F	°C	$°F = \frac{9}{5}°C + 32$ $°C = (°F - 32) \times \frac{5}{9}$
Mass flow rate	lb/h	kg/s	1 lb/h = $1{\cdot}260 \times 10^{-4}$ kg/s 1 kg/s = 7937 lb/h

Volume flow rate	gal/min gal/h	litre/s litre/s	1 gal/min = $7{\cdot}577 \times 10^{-2}$ litre/s 1 gal/h = $1{\cdot}260 \times 10^{-3}$ litre/s
Pressure (operating or working pressure)	lbf/in^2 in H_2O ft H_2O	N/m^2 millibar (mb) mb	1 lbf/in^2 = $6{\cdot}895 \times 10^3$ N/m^2 1 in H_2O = 2·491 mb 1 ft H_2O = 29·89 mb
Pressure drop per unit length	in H_2O/ft	N/m^3 (N/m^{2}m)	1 in H_2O/ft = 817·6 N/m^3 1 in H_2O/10 ft = 81·76 N/m^3 1 in H_2O/100 ft = 8·176 N/m^3
Quantity of heat	Btu	joule (J)	1 Btu = $1{\cdot}055 \times 10^3$ J 1 Btu = 1·055 kilojoule (kJ) 1 therm = 105·5 megajoule (MJ)
Heat flow rate	Btu/h	joule/s = watt (W)	1 Btu/h = 0·2931 J/s or W 1 W = 3·413 Btu/h 1 kilowatt (kW) = 3413 Btu/h
Intensity of heat flow rate	Btu/ft^{2}h	W/m^2	1 Btu/ft^{2}h = 3·155 W/m^2 1 W/m^2 = 0·3170 Btu/ft^{2}h
Thermal conductivity	Btu in/ ft^2 h °F	W/m °C	1 Btu in/ft^2 h °F = 0·1442 W/m °C 1 W/m °C = 6·93 Btu in/ft^2 h °F
Thermal conductance	Btu/ft^2 h °F	W/m^2 °C	1 Btu/ft^2 h °F = 5·678 W/m^2 °C 1 W/m^2 °C = 0·1761 Btu/ft^2 h °F
Specific energy (calorific value)	Btu/lb	J/kg	1 Btu/lb = 2326 J/kg = 2·326 kJ/kg
Calorific value volume basis	Btu/ft^3	J/m^3	1 Btu/ft^3 = $3{\cdot}726 \times 10^4$ J/m^3 = 37·26 kJ/m^3 = $3{\cdot}726 \times 10^{-2}$ MJ/m^3
Heat capacity (volumetric)	Btu/ft^3 h	W/m^3	1 Btu/ft^3 h = 10·35 W/m^3 1 W/m^3 = 0·9661 Btu/ft^3 h
Intensity of heat flow	Btu/ft h	W/m	1 Btu/ft h = 0·9616 W/m 1 W/m = 1·040 Btu/ft h

Schedule B2

Drawing Scales

Domestic heating installation plans and working detail drawings are executed to the following scales

	Imperial scale	*SI scale*
Site plans	1 : 500	1 : 500
	1 : 192 ($\frac{1}{16}$ in to 1 ft)	1 : 200
General scheme	1 : 96 ($\frac{1}{8}$ in to 1 ft)	1 : 100
Working plans	1 : 48 ($\frac{1}{4}$ in to 1 ft)	1 : 50
Details	1 : 24 ($\frac{1}{2}$ in to 1 ft)	1 : 20
	1 : 12 (1 in to 1 ft)	1 : 10
	1 : 4 (3 in to 1 ft)	1 : 5

Schedule B3

Conversion factors: non-SI metric units to equivalent SI units

Pressure

1 mb (millibar)	= 100 N/m^2
1 bar	= 10^5 N/m^2
1 kgf/m^2	= 9·807 N/m^2
1 mm H_2O	= 9·807 N/m^2

Heat

1 cal (calorie) = 4·187 J

Heat Flow Rate

1 cal/s (calorie/second) = 4·187 J/s or W (watts)
1 kcal/h (kilocalorie/hour) = 1·163 J/s or W

Specific Energy
(calorific value)
by weight)

1 kcal/kg	= 4·187 kJ/kg
1 cal/g	= 4·187 kJ/kg

Calorific Value
(Volume basis)

1 kcal/m^3	= 4·187 kJ/m^3

Thermal Capacity per Unit Mass

1 kcal/kg °C	= 4·187 J/g °C
1 cal/g °C	= 4·187 J/g °C

Thermal Capacity per Unit Volume

1 kcal/m^3 °C	= 4·187 kJ/m^3 °C

Intensity of Heat Flow Rate

1 kcal/m^2 h	= 1·163 W/m^2

Thermal Conductance

1 kcal/m^2 h °C	= 1·163 W/m^2 °C

Thermal Conductivity

1 kcal m/m^2 h °C	= 1·163 Wm/m^2 °C

Schedule B4

The following is a list of publications which deal with the change from the Imperial to the SI system of measurement.

Published by the British Standards Institution, Sales Office, Newton House, 101 Pentonville Road, London N1

BS. 350: Part 1: 1959.	Conversion Factors And Tables	£0·75
BS. 350: Part 2: 1962.	Conversion Factors And Tables	£1·25
BS. 350: Part 2: 1962.	Supplement 1967	£1·00
BS. Handbook no. 16.	Miscellaneous Metric Tables	£0·17½
BS. P.D. 5686: 1967.	The Use Of SI Units	£0·10
BS. 1957: 1953.	The Presentation Of Numerical Values	£0·25
BS. 3763: 1964.	International System (SI) Units	£0·30
BS. 6031: 1968.	The Use Of The Metric System In The Construction Industry	£0·40

Published by HMSO, 49 High Holborn, London WC1

'Changing To The Metric System' £0·17½

Published by the Institution of Heating and Ventilating Engineers, at 49 Cadogan Square, London SW1

'Change To Metric' Reference Manual £0·75

IHVE Guide 1970, Book C £6·30

Decimal Currency

Under the Decimal Currency Act 1967, the Decimal Currency Board is responsible for facilitating the transition from the existing currency and coinage to the new system provided for by the act.

The new currency like our present money system is based on the pound sterling, but, whereas we now use three units, the pound (£), the shilling (s) and the penny (d), under the new decimal currency system, two units only will be used, which are, the pound (£) and the new penny, for which the abbreviation will be the letter p, placed after the amount.

The pound is divided into 100 new pence, instead of 240 pence as in our old system and the lowest coin in circulation will be the new halfpenny (½p).

One new penny is thus equal in value to $\frac{240}{100} = 2{\cdot}4\text{d}$, and by conversion:

$$\text{£1 old currency} = \text{£1 decimal currency}$$

$$1 \text{ shilling} = \frac{12\text{d}}{2{\cdot}4} = 5 \text{ new pence (5p)}$$

$$1 \text{ penny} = \frac{1\text{d}}{2{\cdot}4} = 0{\cdot}4167\text{p}$$

For internal accounting, job estimating, estimating running costs and engineering costing generally, it will aid the process of calculation to express the new halfpenny decimally as £0·005, rather than as a vulgar fraction.

In the estimating and accounting fields, decimal fractions of lower value than the ½p must be used (see Chapters 8 and 9).

When pricing goods for sale and in all transactions with the general public, the new halfpenny should be expressed as a vulgar fraction: $\frac{1}{2}$.

The two units mentioned, the pound (£) and the new penny (p) represent the units of account in the decimal system.

Much sound advice and detailed information is contained in booklets prepared by the Decimal Currency Board and Central Office of Information and published by Her Majesty's Stationery Office, two of which are mentioned below:

Decimal Currency: Expression Of Amounts In Printing, Writing And Speech. price 1s (5p)

Decimal Currency: Points For Businessmen. price 1s (5p)

Appendix C
Data on Radiators for Small-Bore Heating

Correction Factors for Other Temperature Differences

Where in makers' lists, emissions are for 56°C temperature difference between the mean radiator temperature and the room air temperature, the emission at any other temperature difference will be equal to

Emission at 56°C (100°F) temperature difference x factor

Temperature difference °C	45	50	56	58	61	64	67	70
Temperature difference °F	80	90	100	105	110	115	120	125
Factor	0·75	0·87	1·00	1·07	1·13	1·20	1·27	1·34

Data in lines 2 and 3 of the above table are quoted from the IHVE Guide 1965.

Effect of Radiator Enclosures

Actual emission = theoretical emission × factor.

Type of enclosure	*Factor*
Radiator in recess with open front	0·9
Radiator in recess with grill in front	0·7–0·8 depending on design of the enclosure

Quoted from IHVE Guide 1965.

Effect of Painting on Radiator Emission*

Ordinary paints and enamels have no appreciable effect on the emission of radiators, and this also applies to the colour of paint.

Aluminium, bronze, and other metallic paints may reduce the overall emission by 10 to 15 per cent.

* From information given in IHVE Guide 1965.

Appendix D
Data on Fuel, Fuel Storage and the Boiler Room

Domestic Oil Fuel Storage

The capacities given below are equivalent to the Imperial values given in *Small Bore Central Heating Guide to Good Practice*

Recommended minimum capacities of fuel-oil storage tanks for domestic central-heating installations

Boiler rating *kW*	*Minimum tank capacity litres*
Up to 18	1250
18 to 26	1400
26 to 35	1800
35 to 44	2700

Average weights and volumes of solid fuel

Fuel	*Volume* $m^3/tonne$	*Weight* kg/m^3
Anthracite	1·27	785
Gas coke	2·61	384
Rexco; Burnbrite; Sebrite and similar fuels	2·08	481
Hard coke	2·15	465
Coalite	2·40	416
Bituminous coal	1·33	753

The above are average values. The true space required by solid fuel depends upon the type and size of the particular fuel.

Area in square metres per tonne of fuel (average) for a stacking height of 1·25 m

Fuel	*Fuel store floor area per tonne* m^2
Anthracite	1·04
Gas coke	2·14
Rexco; Burnbrite; Sebrite and similar types of fuel	1·76
Hard coke	1·71
Coalite	1·97
Bituminous coal	1·09

Solid-fuel sizes for small central-heating and hot-water supply boilers

Type of boiler	*Fuel size*
Hand-fired independent boiler, burning smokeless fuels.	Gas coke no. 3 : 13 mm–32 mm (½ in–1¼ in) Anthracite : 22 mm–41 mm (⅞ in–1 in) Manufactured brands of smokeless fuels in the range : 13 mm–38 mm (½ in–1½ in)
Automatic boiler burning small bituminous coal.	Washed singles : 16 mm–25 mm (⅝ in–1 in)
Automatic gravity-feed, or hopper-type boilers, burning smokeless fuels.	Anthracite beans : 10 mm–22 mm (⅜ in–⅞ in) Manufactured smokeless fuels: 10 mm–19 mm (⅜ in–¾ in) Hard coke : 10 mm–16 mm (⅜ in–⅝ in) Subject to the boiler manufacturer's advice, certain makes of boiler will burn:- Anthracite grains: 5 mm–13 mm (5/16 in–½ in) and peas : 10 mm–16 mm (⅜ in–⅝ in)

Air inlet to boiler room for solid-fuel or oil-burning equipment

Boiler rating kW	*Free area of fresh air inlet grill m^2*
Up to 15	0·01
15–22·5	0·015
22·5–30	0·02

Note: Installers of domestic central-heating equipment using fuel oil should always consult the Local Authority in whose area the proposed installation will be sited, in order to ensure that the proposals comply with the Local Authority's regulations covering the storage and use of fuel oil.

The Importance of Good Ventilation in any Room where a Gas Boiler is Fixed

Between 0·17 and 0·23 m^3 of air are needed for each cubic metre of gas burned in the boiler. Good ventilation of the room in which the boiler is placed is therefore necessary to ensure a good air supply to the burner.

An air inlet placed approximately 300 mm above floor level should have a minimum free area of 0·07 m^2, or twice the cross-sectional area of the flue, whichever is the larger. High-level air outlets placed above the boiler, should have not less than two-thirds the free area of the fresh-air inlets.

To ensure adequate air circulation at the gas burner level in basement boiler rooms, air inlets must be extended in ductwork to discharge near the floor.

Appendix E
Data on Pipes and Equipment

Copper tubes for domestic central-heating and hot-water supply (to BS 659: 1955)

Nominal bore		*Gauge SWG*	*Water content*		*Weight*	
mm	*in*		*litre/m*	*gal/ft*	*kg/m*	*lb/ft*
6	$\frac{1}{4}$	20	0·045	0·003	0·201	0·135
10	$\frac{3}{8}$	20	0·075	0·005	0·283	0·19
13	$\frac{1}{2}$	19	0·134	0·009	0·402	0·27
19	$\frac{3}{4}$	19	0·283	0·019	0·580	0·39
25	1	18	0·522	0·035	0·923	0·62
32	$1\frac{1}{4}$	18	0·805	0·054	1·131	0·76

Imperial values quoted from IHVE Guide 1965.

Steel tubes for domestic central heating (to BS 1387: 1957) (medium grade)

Nominal bore		*Gauge SWG*	*Water content*		*Weight*	
mm	*in*		*litre/m*	*gal/ft*	*kg/m*	*lb/ft*
10	$\frac{3}{8}$	13	0·119	0·008	0·853	0·573
13	$\frac{1}{2}$	12	0·208	0·014	1·223	0·822
19	$\frac{3}{4}$	12	0·373	0·025	1·577	1·060
25	1	10	0·582	0·039	2·440	1·640
32	$1\frac{1}{4}$	10	1·014	0·068	3·140	2·110

Imperial values quoted from IHVE Guide 1965.

Copper pipes: Intervals between pipe supports

Nominal bore of pipe mm	*Interval for vertical runs m*	*Interval for horizontal runs m*
6	1·25	1·00
10	2·00	1·25
13	2·00	1·25
19	2·50	2·00
25	2·50	2·00
32	3·00	2·50

Steel pipes: Intervals between pipe supports

Nominal bore of pipe mm	*Interval for vertical runs m*	*Interval for horizontal runs m*
10	2·00	1·25
13	2·50	2·00
19	3·00	2·50
25	3·00	2·50
32	3·00	2·50

Note on Future Metric Pipe Sizes

At the time of writing, plans for the production of a full range of metric-sized tubes and fittings in copper, steel and other materials have not been finalised.

Even after the complete changeover is accomplished, the data on pipes, fittings and pipe sizing in this and similar books will continue to be useful when dealing with modifications and extensions to existing systems.

Standard sizes of open cisterns in common use for domestic central heating, and hot-water supply (galvanized steel)

Nominal capacity gal	*BS no*	*Actual capacity gal*	*Length, width and height in.*
10	C.1	4	18 × 12 × 12
15	C.2	8	24 × 12 × 15
20	C.3	12	24 × 16 × 15
25	C.4	15	24 × 17 × 17
30	C.5	19	24 × 18 × 19
40	C.6	25	27 × 20 × 20
40	C.7	25	24 × 24 × 19
50	C.8	35	29 × 22 × 22
60	C.9	42	30 × 23 × 24

Note: The minimum thickness of steel sheet before galvanising, Grade A. 16 BG., Grade B. 18 BG.
Extracted from BS 417: 1951, by permission of the British Standards Institution, from whom copies are obtainable at Newton House, 101 Pentonville Road, London N1.

Standard sizes of open cisterns in common use for domestic central heating and hot-water supply. This table is based on the Imperial data in the previous table, rounded off to the nearest whole number

Nominal capacity litre	*Actual capacity litre*	*Length, width and height mm*
45	18	457 × 305 × 305
68	36	610 × 305 × 381
91	55	610 × 406 × 381
114	68	610 × 432 × 432
136	86	610 × 457 × 483
182	114	686 × 508 × 508
182	114	610 × 610 × 483
227	159	737 × 559 × 559
273	191	762 × 584 × 610

Conversion factors: 1 gal = 4·546 litre; 1 in = 25·4 mm.

Standard metric sizes for copper indirect cylinders used with combined central heating and hot-water supply in small dwellings

BS type reference	*Min. storage capacity*	*External dimensions mm*		*Size of pipe connections*	
		Diameter	*Height*	*Cylinder mm*	*Primary heater mm*
1	114	410	1070	25	25
2	100	460	760	25	25
3	123	460	910	25	25
4	145	460	1070	32	32
5	170	460	1210	32	32
6	214	510	1270	38	38

Extracted from BS 1566: Part 1: 1966 by permission of the British Standards Institution, from whom copies are obtainable at Newton House, 101 Pentonville Road, London N1.

Standard Imperial sizes for copper indirect cylinders used with combined central heating and hot-water supply in small dwellings

BS type reference	*Min. storage capacity*	*External dimensions in*		*Size of pipe connections*	
				Cylinder	*Primary heater*
	gal	*Diameter*	*Height*	*in*	*in*
1	25	16	42	1	1
2	22	18	30	1	1
3	27	18	36	1	1
4	32	18	42	$1\frac{1}{4}$	$1\frac{1}{4}$
5	37	18	48	$1\frac{1}{4}$	$1\frac{1}{4}$
6	47	20	50	$1\frac{1}{2}$	$1\frac{1}{2}$

Extracted from BS 1566: Part 1: 1966 by permission of the British Standards Institution, from whom copies are obtainable at Newton House, 101 Pentonville Road, London N1.

Appendix F
Insulation

Minimum thickness of insulation for low-pressure hot-water central heating and hot-water supply systems in small dwelling houses, applicable to plant and pipework

Nominal pipe bore mm	*Thermal conductivity k of the insulating material (Wm/m² °C)*				
	0·0288–0·0433	0·0447–0·0606	0·0620–0·0692	0·0707–0·0793	0·0808–0·0865
	Minimum thickness of insulation, mm				
Up to and including 25	13	19	25	32	38
Above 25 and not exceeding 76	19	25	32	38	44
Above 76 and flat surfaces	25	32	38	44	50

Insulation of the Building Structure

Modern methods of insulating the roof, walls and floors, together with double glazing of windows will, if carried out to comply with the recommendations of the Egerton Committee (1946), produce savings in annual costs which will quickly repay the capital cost of insulation.

The Egerton Committee recommended structural insulation to give the *U* values shown in the following table:

Structural insulation: recommended U values

Construction	*Recommended U value* $W/m^2\ °C$
External walls of living rooms	0·85 to 1·14
External wall of other rooms	1·14 to 1·70
Ground floor	0·85
Roof and top floor ceiling	1·14 to 1·70

Some constructions that conform to the officially recommended standards of insulation

Construction	*U value* $W/m^2\ °C$
External Walls	
279 mm (11 in) sealed cavity brick wall, plastered internally, with 25 mm (1 in) glass-fibre slab (or other slab of equal 'k' value) fitted to the inner face of the wall cavity	0·79
Cavity of the above wall filled with urea-formaldehyde foam	0·62
Ground Floors	
Solid concrete floor on ground with 25 mm (1 in) expanded plastic below screed or slab	0·62
25 mm (1 in) T and G flooring on timber joists, and ventilated beneath, with 25 mm (1 in) expanded plastic (or other material of equal 'k' value) between the joists on holding nails	0·79

Pitched Roof	
Tiles on battens, on sarking felt, plaster or plaster-board ceiling, with 25 mm (1 in) to 38 mm ($1\frac{1}{2}$ in) glass fibre; granulated cork; mineral wool; or vermiculite, loose fill	0·62
Flat Roof	
102 mm (4 in) reinforced concrete, covered with asphalt or roofing felt, plastered soffit, with 25 mm (1 in) expanded plastic (or other suitable material of equal 'k' value) under roofing felt, or as permanent shuttering	0·91

Index